The Storied Places of West Maui

The Storied Places of West Maui

History, Legends, and Place Names of the Sunset Side of Maui

Michelle Anderson

From the research of Maui County Historian Emeritus

Inez MacPhee Ashdown

North Beach-West Maui Benefit Fund Inc.

Lahaina, Maui, Hawai'i

Printed in the United States of America

First printing, 2016
Second printing, 2023

ISBN 978-0-8248-6734-8 (pbk : alk. paper)

Published by the North Beach-West Maui Benefit Fund, Inc.
P O Box 11329
Lahaina, Hawai'i 96761

Distributed by University of Hawai'i Press
2840 Kolowalu Street
Honolulu, HI 96822-1888

This book is printed on acid-free paper and meets the guidelines for permanence and durability of the Council on Library Resources.

Print-ready files provided by the author.

To Tūtū ʻĀina Kaulana,
and all the Keiki o Ka ʻĀina of Maui.

CONTENTS

LIST OF FIGURES

All photos without attribution are from the author's collection.

ACKNOWLEDGMENTS

I would first like to thank the late Inez Ashdown, Maui County Historian Emeritus. Her inspiration motivated me to write this book for her.

Mahalo nui loa to the *kūpuna* (elders) of the West Maui Hawaiian Civic Club for embracing Mrs. Ashdown in the 1930s as one of their own and sharing their heritage with her. We are also indebted to her *kahu* Kīnaʻu, the guardian cowboy of *aliʻi* descent who inspired her love of the *moʻolelo* (oral history) of his people at a very young age.

A generous thank you to the North Beach-West Maui Benefit Fund, Inc. for their support in the research and writing of this book, and for its publication. Profits from this book will go into a perpetual publication fund to keep the book in print for future generations. *Mahalo nui loa* to the Board of the North Beach-West Maui Benefit Fund for this commitment to the *keiki o ka ʻāina* (children of the land) of Maui. A special thank you to Lance Collins, PhD, for his guidance through this long process.

A very special thank you to the staff and Board of the Maui Historical Society at the Bailey House Museum in Wailuku for their generous assistance in providing me access to over 500 selected pages of research from Mrs. Ashdown's archive.

I would like to thank all those who generously permitted the use of their stunning photographs to illustrate the stories in this book.

A grateful *mahalo* to my dear friend Clare Apana, who graciously hosted me in her home on my research trips to Maui.

A heartfelt *mahalo* to my friend Kathy Kaohu for her assistance with copying the many documents from the Maui Historical Society and for accompanying me on my "photo safaris" of West Maui.

And last, but not least, to my sister Candace "Pilialoha" Shaffer, a big thank you for her wise advice, encouragement, and critical review of the manuscript.

INTRODUCTION

The Storied Places of West Maui lay hidden in the place names given to its mountains and streams, its valleys and springs, its ancient fishponds and stone temples, and even large *pōhaku* (stones). Every name has a special meaning and many names have *moʻolelo* (stories) attached to their translations, making them *Wahi Pana,* Storied Places. The stories recall important events, important people, and the spiritual dimensions of an ancient culture, all found in the translation of a name given to a particular place very long ago.

When these storied places were named and by whom is unknown for the most part, but all of the ruling dynasties of Maui, from the Hua Dynasty of the 10th century to the Kekaulike Dynasty of the 18th century, are represented in the *wahi pana* of West Maui. Hawaiʻi-loa, who arrived long before ruling dynasties were formed, is attributed with naming the eight major islands and the eight sea channels of Hawaiʻi, along with many landmarks on Maui and the island of Hawaiʻi.

In the early 19th century Lahaina became the capital of the new Hawaiian Kingdom under Kamehameha I and was at the epicenter of Western impact that would rapidly change the ancient ways forever, replacing them with Christianity and a westernized form of government and commerce. The *wahi pana* of this important transitional era in Lahaina are also included.

These *wahi pana* contain the history, both mythological and historic, of the ruling dynasties of West Maui and their gods and goddesses. Passed on in the ancient oral traditions of *oli* (chants) and *moʻolelo* (stories) in the courts of the *aliʻi* (chiefs), many *wahi pana* were considered sacred and were not told in the dialect of the *makaʻāina-na* (commoner). Early Hawaiians spoke in three dialects. Because arcane knowledge was considered sacred and not to be misused or misinterpreted by the uninitiated, the *aliʻi* and their *kāhuna* (priests) spoke in the *hūnā* (secret or hidden) dialect, which hides its deeper meanings in parables.[1] Often of higher thoughts and spiritual things, the true meaning of a place name is usually hidden in the *hūnā* translation.[2]

The *haku mele* (poets, bards) spoke in the dialect called *ka-ona*, the poetic, double-meaning sounds of imagery used to "spice-up" long narratives, such as the *Kumulipo,* a two thousand-line Creation Chant.

These esoteric dialects died out long ago, but some place name translations in the *hūnā* dialect have survived through the descendants of their *kūpuna aliʻi* (chiefly elders).

Inez MacPhee Ashdown, 1982

Local historian Inez MacPhee Ashdown began writing publicly about the Hawaiian culture for the *Honolulu-Star Bulletin* and the *Maui News* in 1936. Her first article was about *kumu hula* (dance teacher) Emma Sharpe's *hula ʻūniki* (graduation ceremony) and a brief history of the hula. After writing a chant honoring the famed hula dancer, Iolani Luahine, the Maui Hawaiian Women's Club gave Inez the name ʻĀina Kaulana, engraved on a gold Hawaiian bracelet. It meant she brought "Pride to the Land."

After the United States took control of Hawaiʻi in 1898, the Hawaiian language, history, and culture had been suppressed. Emphasis in the schools was on becoming "American" and to "swim with the overwhelming tide." Inspired by meeting Queen Liliʻuokalani as a child of eight and having learned about the Hawaiian culture since 1908 from the Hawaiian *kahu* (guardians) of her youth, Inez was dedicated to the preservation of the culture. Her stories were well received and she was invited as a guest speaker to many organizations where she began advocating for the saving of historic sites and the *wahi pana* (storied places) of the island and for the teaching of the Hawaiian language and genealogies. Soon, Mrs. Ashdown was invited to become an honorary member of the West Maui Hawaiian Civic Club (WMHCC) and the East Maui Hawaiian Civic Club, where historic sites committees were set up under her direction to begin documenting the storied places of their districts.

Many of the *kūpuna* (elders) in the West Maui Hawaiian Civic Club (WMHCC) were charter members from old *ali'i* families. They still remembered the *mo'olelo* (stories) of their elders and wanted it preserved before it was too late. These noble Hawaiian *kūpuna* banded together, at a time when the history of their own culture was still not taught in Hawai'i's schools, and began identifying the storied places of their districts and Mrs. Ashdown began writing it down.

The matriarchs of the West Maui Hawaiian Civic Club were the Shaw sisters, Alice Ka'ehu-kai Shaw Ka'ae and Mary Kawai'ele Shaw Hoapili. Born on Moloka'i in the mid-1860s, they were adopted by King Lot Kamehameha V who gave them the names Ka'ehu-kai (The Sea Mist) and Ka'wai'ele (The Deep Waters). From *ali'i* heritage, their grandfather Nalehu was a grandson of Hawai'i's King Alapa'inui and their birth father, Kanaina, was a relative of King Lunalilo. Their mother, Lahela Kuahu'ula Nalehu, was born in Lua'ehu, Lahaina in 1832. She later married Patrick Shaw, Governor of Moloka'i under Kamehameha V.[3]

Alice Ka'ehu-kai Shaw Ka'ae, Aunty Kai

Witness to the last days of the monarchy, the Shaw sisters lived at the center of Lahaina's ancient past, in the royal enclave of Lua'ehu next to the Wai-o-kama fishpond. Both sisters had served as Ladies-in-Waiting to Hawai'i's last monarch, Queen Lili'uokalani. Their knowledge of Maui *ali'i* and the storied places of their region was an invaluable resource that they freely shared with Mrs. Ashdown and the historic sites committee. Aunty Kai (for Ka'ehu-kai) and Aunty Ka'wai'ele would lead Inez and other elders on expeditions up and down the coastline of West Maui. Aunty Kai would say "Halt," and Inez would pull over and write down everything she said about the place.[4]

Other elders of chiefly descent on the historic sites committee were: Annie Hinau, whose mother Ka-ohu-lani was also a Lady-in-Waiting to Queen Lili'uokalani, and Annie's husband Junior Hinau, who was raised in his grandfather *Napaepae*'s house next to the Kings Palace in Lua'ehu,[5] Alice Makekau Aki, who attended Hale Aloha and the Lua'ehu school as a child [her mother, Mima Makekau, was of the priestly class of *ka-huna lapa-au* (natural healers)],[6] Annie Ako, Lahela Reimann, Hattie Kanaka-o-kai Yoshikawa and her sister Elizabeth Namau'u, Mary Chan Wa, William Saffrey, Robert Saffrey, Winnie Saffrey Sanborn, Anaka-luahine, John Lihau and his brother George Ka'aehui, Tommy Kekona, John Kapa-ku, Reverend John Kukahiku, Pilahi Paki, Kaniho Humehume, Sam Makekau, James Kahahane, and Manuel Silva, were all contributing

Hattie Kanaka-o-kai Yoshikawa and Inez at Kaupō

members, adding to the knowledge of their specific region of West Maui. The committee discussed the stories and Mrs. Ashdown recorded them.[7]

Many of West Maui's storied places were rediscovered and identified by the WMHCC during the 1930s and 40s. Also, long before there was any official system of historic preservation, through their lobbying efforts, the WMHCC was able to enlist the assistance of local government and the plantations and saved the original missionary home, Baldwin House, the original printing house, Hale Pa'i, and the original prison, Hale Pa'aho, from total collapse. These original buildings are now the crown jewels of the Lahaina Historic District.[8]

The elders of the WMHCC had accomplished a lot as volunteers, but after World War II they wanted a more comprehensive approach to the preservation of their historic sites and they wanted the Territorial Government to participate. In 1950 they invited a group of legislators to Lahaina to plead their case. Their plea was heard and in 1951, the Territorial Historic Sites Commission was created under Governor Long.[9]

Mrs. Ashdown was appointed as the Historic Sites Commissioner for Maui County and began working with Hawaiians across the county to search out sites and inventory them with modern tax map key numbers for permanent identification. Many sites were saved by the Maui Historic Sites Commission, including the *heiau* (stone temples) Haleki'i and Pi'ihana in Wailuku, Pi'ilani Hale in Hana, and Wailea Village at Palauea. Pioneer Mill, under manager John Moir, a member of the Historic Sites Commission, agreed to set aside for preservation the Heiau Heki'i of 'Ukumehame Valley, the Heiau Ka'iwaloa in Olowalu, and Pu'u Kilea with its petroglyphs, back in 1956, unheard of for a company to do at the time.[10]

In 1968 Mrs. Ashdown was hired as County Historian to continue locating and identifying historic sites with the group Hui Hana Malama (Group Working to Protect). In 1970, she wrote her book, *Ke Alaloa O Maui* (The Broad Highway of Maui) to document the work of the Hui.[11]

In 1975 she left her field work and began researching and writing "The Districts of Maui" for the county. Mayor Cravalho wanted the names of beaches, streams, and valleys and the history of every district of Maui written down. For Mrs. Ashdown, that included the storied places of the district. For West Maui she used the stories shared with her from the *kūpuna ali'i* of the West Maui Hawaiian Civic Club. These stories are the primary source material for the storied places shared in this book.

Retiring in 1978, Mrs. Ashdown left an archive of her work for the benefit of future generations, "to know their rich heritage." In 1999 her exhaustive body of work was finally catalogued and indexed at the Maui Historical Society and is available to researchers at Bailey House Museum in Wailuku. It is from this archive, as well as a personal collection of materials given to me by Mrs. Ashdown, that this book was researched and written.

Mrs. Ashdown with John Pi'ilani of Hāna. Together with William DeRego, they saved Heiau Pi'ilani Hale.

The history surrounding these storied places has been included for historical context, but is not meant to be a definitive history of West Maui. See the Bibliography for excellent resources. Complex Hawaiian words have been hyphenated the first time they are used to assist in proper pronunciation.

Just as places are regional, so are the place names and the oral traditions attached to them. Words and translations vary according to the area, social background, and knowledge of the old people of long ago. Early writers did not usually know anything but the "every day" language of the *maka'āi-nana* (commoner) called *ho'o-pu-kaku*, so the deeper meanings of place names have often been lost. Also, the slightest misuse or mispronunciation of a word can change its meaning entirely, which often happened with the passing of time.[12] Additionally, many words have several common meanings, depending on the context used.

Some *kūpuna ali'i*, like Aunty Kai, still knew the *hūnā* translations of many *wahi pana* (storied places), unlocking their deeper meanings. But the esoteric translations of many place names have been lost to time and only *ho'opukaku* (literal) meanings remain, if we are lucky.

The Reverend Edward Kapo'o and the Reverend Moses Kahai'apo, two of the last on Maui who knew the *ka'ona* and *hūnā* dialects, worked with Mrs. Ashdown to translate some important names in this manner.[13]

In 1982, Mrs. Ashdown was honored with the title of Maui County Historian Emeritus by Mayor Hannibal Tavares for her years of contribution to historic preservation. For more on Mrs. Ashdown's extraordinary life and her leadership role in preserving Maui's past, see the Appendix: 'Āina Kaulana.

Archeologist Winslow Walker surveyed the *heiau* of Maui in 1928 and 1929. In his 1931 report, "Archeology of Maui," he offers a chronology of Maui *ali'i* based on the early work of Judge Abraham Fornander's *An Account of the Polynesian Race*, published in 1879, and follows the Ulu-Hema genealogy of the Maui *ali'i*. Fornander's work suggested that the first migration to Hawai'i happened between 400 and 600 AD from the Marquesas and a second migration from Tahiti began in the 11th century, which established the ancient Hawaiian culture. Scholars still debate these dates.

While Walker's chronology of the Ulu-Hema genealogical line of Maui Ali'i was only "suggestive," it appears to have been used by many subsequent historians. Mrs. Ashdown's dates generally follow Walker's chronology, which only goes back as far as Paumakua (975 AD), because according to Walker, it "is impossible to separate the historical characters from the mythical cosmological ones with accuracy."[14]

If you are not native Hawaiian, you may have to suspend your belief system in order to fully appreciate the mythology found in many of these storied places, when god-men walked the earth. Because most storied places were named so long ago, they are imbued with both the mythological and the historic past of Hawai'i, that at times seem to intersect.

For those unfamiliar with the Hawaiian culture, the opening chapter, *A Divine Past*, offers a brief overview of the ancient belief system, beginning with the Hawaiian Story of Creation, the *Kumulipo*. It is hoped that this small offering will provide a deeper appreciation for the following stories found in the place names of West Maui. For those who are *kānaka* (Hawaiian), my humble apologies for the very brief explanation of a highly complex belief system.

The *mo'olelo* (oral history) attached to the ancient names of West Maui are told by the last generation of *kūpuna ali'i* who were alive during the Hawaiian monarchy. It was in the courts of the *ali'i* where the *haku mele* (poets and bards) shared the knowledge of their divine past. We are forever indebted to the *kūpuna ali'i* of the West Maui Hawaiian Civic Club for sharing their family stories with Mrs. Ashdown, who they trusted, and to Mrs. Ashdown for her dedicated efforts to get them in print for future generations.

Come, walk in the footsteps of the *ali'i*, along the face of Maui, from 'Uku-me-ha-me to Kaha-ku-loa, and find the stories they wanted told in the *wahi pana* (storied places) they left behind.

A DIVINE PAST

Olowalu Petroglyph

The Hawaiian Story of Creation begins in deep darkness and tells a poetic epic of the creation of the earth and all living things. Preserved in the *Kumulipo,* an oral chant over 2,000 lines long, it tells of the First Era of Creation where all lower life forms were born, beginning with the smallest creatures of the sea, to the Second Era of Creation and the coming of the gods and mankind. It tells of the many manifestations of Ke Akua (The One God) sent to shape the world: as Kāne (The Creator), as Lono (The Sustainer of Life), and as Kanaloa (the Ruler of Death and Eternal Life), all sent to prepare the earth for mankind.

It tells the story of Kāne and Hina-li'i of the Heavens, parents of the Hi'iaka Family of the Elements, demi-gods born to assist in the formation of the universe. Their daughter, Pele, the Fire Goddess of Volcanos, shaped and formed the land. Her brothers and sisters, the many sea, sky and air gods and goddesses, ruled all of the natural elements. Many generations of demi-gods and their deeds are listed in the *Kumulipo* chant, including Māui Lele, or Māui-a-ka-lana, famous throughout all Polynesia for his many super-human deeds, including slowing the sun for his mother Hina.

Kāne and Kanaloa were the first *akua* (gods) to walk upon the Earth, creating fresh water springs across the land.[15] Then Kāne shaped a figure like himself in the sacred red earth and breathed his spirit, his *Hā*, (the Breath of Life) into it. A mortal man, Ke Ali'i Kū Honua, The Lesser God (Ke Ali'i), who Stood Upright (Kū) from the Earth (Honua), was born. Kāne then gave him a mate, Ke Ali'i Lalo Honua, The Lesser God of the Supine Earth, and mankind was born as god-men of divine origin.

Kāne then walked among the first men and women and taught them the *Kānā-wai Akua* (Laws of God). Lono was the patron of all plant life and taught mankind how to farm and irrigate and the proper use of medicinal herbs. Kanaloa was in charge of the streams flowing to the ocean and the Stream of Life, flowing into the Sea of Eternity beyond the western horizon.[16]

Distinguished descendants of these god-men were often deified at death to become *ʻau-ma-kua,* Ancestral Guardian Spirits, for their families. They took many forms including the *pu-e-o* (owl), the *mano* (shark), and the *moʻo* (lizards or dragons). With the supernatural power to physically manifest wherever needed, many *ʻaumakua* became famous in legend for heroic deeds on behalf of their ancestors. The *aliʻi* families of Maui had the famed Kiha Family of *moʻo* goddesses as their family *ʻaumakua.* These fresh water spirits lived in lakes and ponds and streams, and could take the form of a giant lizard or dragon, or even a beautiful mermaid.

From the beginning, West Maui was home to the highest born *aliʻi*, where the purest-bred of all could trace their ancestry straight back to Kāne. Rank and power in ancient Hawaiʻi came through divine descent from the gods, memorialized in the *Moʻo-kū-ʻau-hau,* the lengthy genealogies of *aliʻi* dynasties. Most genealogies began with the stages of creation to the first *aliʻi* from whom the chief could claim descent and on through the many generations to himself. There were many genealogies known by the ancient chanters, such as the *Kumu-ho-nua* and *Kumu-ʻulu* traditions,

Boki, Governor of Wahu of the Sandwich Islands, and his wife, Liliha.
Printed by C. Hullmandel; drawn on stone from the original painting by John Hayter. London 1824. Ref: C-020-009. Alexander Turnbull Library, Wellington, New Zealand

following different branches of divine *ali'i*. The best known *Mo'okū'auhau* (genealogical chant) in modern times is the *Kumu-lipo*, the creation chant that begins in deep darkness. [17] (*Kumu* – root, origin; *lipo* – deep blue black of the sea.)

The *Kumu'ulu* tradition begins with Kumuhonua, the first man, and Lalo-honua the first woman, down to Nu'u of the 12th generation, who survived the Great Deluge, then on down to Hawai'i-loa Ke-kōwā-i-Hawai'i, The Great Discoverer. About thirty-seven generations from Kumuhonua come Wākea and Haumea-Papa. From there, the Maui *ali'i* followed the 'Ulu-Hema Genealogy.[18]

These genealogical chants were carefully preserved by artists trained in the memorization and exacting rendition of the sacred histories. It is because of the *mo'okū'auhau* (genealogical chants), preserved through the strict oral tradition of ancient Hawai'i, that today we know the names and deeds of the historical and mythological ancestors of Hawai'i's ancient past.[19]

Marriage among the *ali'i* was often arranged in an effort to keep divine bloodlines pure and also to secure alliances among chiefs. Parents arranged these betrothals when the children were very young. The union of a brother and sister of rank was not unusual in order to intensify status and *mana* (spiritual power). The higher the rank, the greater the *mana* the individual possessed. Called a *pi-'o* (the arch of a rainbow) marriage, the offspring of this union were called *nī'au pi'o*, or Divine Being. Maui's great Queen Ke'ōpūolani was the last of the *nī'au pi'o ali'i*, the highest ranking of all *ali'i* when she married Kamehameha I, producing heirs whose divine rank, and therefore right to rule, could not be questioned. So sacred was Ke'ōpūolani, that those in her presence had to bare their chests and prostrate themselves, including her husband, the King.[20]

As rulers by divine rank, the *ali'i nui* possessed great *mana* and the supreme power of life and death. This power was protected by a complex and fearful *kapu* (taboo) system of laws and restrictions that permeated every aspect of life for both the commoner and the ruling chiefs. Believed to be decreed by the gods and interpreted by the *ali'i*, these laws of conduct were so strict that a broken *kapu* meant almost certain death. Only the Maui *ali'i* possessed the power of the Kapu of the Burning Sun, which meant death by burning.

Each island had places of refuge called *pu'u-ho-nua* where offenders could find forgiveness from the attending priest if they could reach it before capture.

From the earliest *ali'i*, who walked with the gods and goddessess, to the many ruling dynasties of Maui, their stories are told in the *Wahi Pana* (Storied Places) they left behind.

LAHA'ĀINA-LOA
Broad Land of Prophesy

All of West Maui was once known as Laha'āina-loa. Photo by Ivansabo at Dreamstime.com

In very ancient times, all of West Maui was called Laha'āina-loa, Broad Land of Prophesy, where the sacred *ali'i* who descended from The Beginning lived and had direct contact with the Creator and his god-spirits.[21] Laha'āina-iki was all the rest of Maui.

These first *ali'i* were prophets, or seers, with the *mana* (spiritual power) of "second sight." They proclaimed the *Kānāwai Akua* (Laws of God) to the *kāhuna* (priests) for instruction to the populace. One of the earliest prophets was Laha'āina-loa, who gave his name, Broad Land of Prophesy, to all of West Maui. Eventually Laha'āina became known as the place of "coming and going," Lele, the center of West Maui and center of the island chain, where many canoes came and went from the languid shores of Laha'āina, Land of Prophesy.[22]

Maui's earliest known dynasty of rulers, the Hua Family, came from the south and ruled all of West Maui and Hāna by at least 1100 AD. Many of the stone temples of West Maui are attributed to them. *Heiau* were temples of worship and of learning under the guidance of *kāhuna* (priests) who held the secrets, the *hūnā*, passed on from generation to generation.

The platform-type *heiau* have commanding views with stone filled platforms built outwards toward the sea from an existing place of high ground. Often the face of the platform is terraced in stone. Other types are walled enclosures built from ground level. Constructed centuries ago without the use of mortar, those few heiau that have survived undisturbed still stand with the integrity of an unforgotten name passed on through centuries of generations. Some are still storied places, where ancient history and legends live on.

MAUNA KAHĀLĀWAI

Mauna Kahālāwai. Photo by Phillip Colla

West Maui is formed by a range of mountains called Mauna Ka-hālā-wai, Mountain Blessed with Waters. Symbolically, it is the Juncture between Heaven and Earth, where the Twin Waters of Life meet: rain from Wākea (Sky Father) and springs from Haumea-Papa (Earth Mother).

PUʻU KUKUI

Puʻu Kukui. Photo by Ron Dahlquist

Puʻu Ku-kui, Hill of Light and Enlightenment, is the highest peak of Mauna Kahālāwai at 6,070 feet, forming the apex of the ridges named Kahoʻolewa, Lua-ko'i, and Na-kalu-lua. Pu'u Kukui is said to be the third rainiest spot on earth with an average annual rainfall of 400 inches. From this peak, the deep valleys and streams of West Maui radiate, like the spokes of a wheel, giving Maui the state nickname of "The Valley Isle." Although West Maui has always been hot and arid, its abundant stream waters from Mauna Kahālāwai provided for the cultivation of farmland, *lo'i kalo* (taro patches) and *loko i'a* (fishponds) throughout the region.[23]

LAKE MANOWAI

Photo by Forest and Kim Starr

Within Pu'u Kukui lies Lake Man-o-wai, which means Source of Life. It is said that the gods Kāne and Kanaloa first struck their spears here to create the waters of Manowai. Because of this, the outlet of the lake is called 'Ō-maka, or Beginning.

From Lake Manowai flow the headwaters and underground springs that feed the stream waters of West Maui.[24] Growing in Lake Manowai are rare silverswords, cousins to those on Haleakala, called Lau 'Olo-ke-le, or bog leaves, because they live in water.

Lake Manowai was also known as Mano-wai o Kiha, or Ki-'o-wai o Kiha, in reference to the many Kiha family spirits who have lived here. Said as *mano,* with no accent on the syllables, it means numerous. Said as *ma-no* it means source of water (*o-wai*) or life. [25] In the early 1900s the manager of Honolua Ranch discovered rare violets here and renamed the pond Violet Lake, as it is still known on maps today.[26]

MAUNA ʻEʻEKE

Photo by Forest and Kim Starr

North of Puʻu Kukui stands Mauna ʻEʻeke at 4,480 feet. This flat-topped crater juts up from the mountain several hundred feet providing isolation for its rare and endemic plants. It once was the home of the goddess Pele, where she retreated to avoid her water relatives on her journey across Maui. Now, Mauna ʻEʻeke provides headwaters for Ka-haku-loa and Hono-kō-hau streams.[27]

ʻEʻeke (now spelled ʻEke on maps) is the proper name and means the shrinking, the elusive, retreating mountain, hidden in the clouds. Normally shrouded in the mists, it was said that when ʻEʻeke is hidden, the Hiʻiaka Family of Elements and phenonmena are at odds and the priests who could "read the omens" of the weather could predict when to work, when to fish, how to plan.[28]

A Maui legend tells that Nuʻu and his wife, Nana-nuʻu, landed their canoe on ʻEʻeke after the Great Flood. From there Nuʻu took his family to the eastern end of the island and established the village of Nuʻu. It is said that Lake Manowai became sacred because it held the last waters of the Kai-o-Hīnaliʻi (the Sea of Hīnaliʻi), or the Great Deluge, that covered the continent of Mu and left its mountaintops as the islands of the Pacific. At Puʻu Kukui, Hill of Enlightenment, *Ke Akua* (God) told Nuʻu that the rainbow would always be a sign of promise to mankind. After Nuʻu and Nana-nuʻu died, their children buried them on ʻEʻeke. [29] Nuʻu appears in the twelth generation of the Kumuhonua genealogy.[30,31]

HAWAIʻI LOA KE KŌWĀ

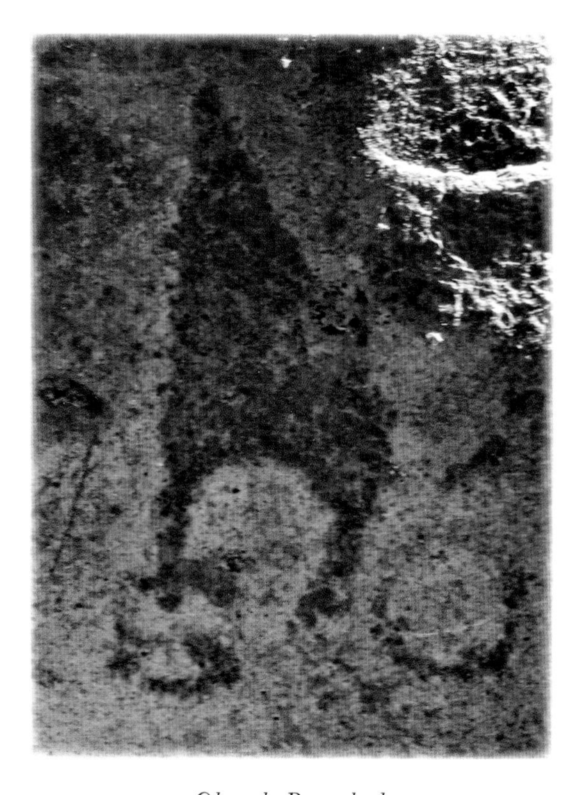

Olowalu Petroglyph

Long after the Great Flood, perhaps as far back as 88 BC, the legendary Hawaiʻi Loa Ke i Kōwā (Hawaiʻi of the Sea Channels) discovered these islands. Traveling across the great seas from Kahiki Kū (the Great Horizon), with only the stars to guide them, Hawaiʻi Loa and his helmsman Makaliʻi (Eyes of the Chief) steered their canoes into the most isolated group of islands in the Pacific and claimed it for settlement. Following Pleiades and Jupiter, the guide stars in the fall of the year, Hawaiʻi Loa and his crew first saw land covered in forests of wili-wili trees, blanketed with seeds flaming orange like a beacon, and so he called this island Kohema-lama-lama, or Southern Beacon.[32] Other legends also named the island Kana-loa, for the diety who made fresh springs throughout Hawaiʻi with his brother Kāne. Eventually, Kohemalamalama was named Kahoʻolawe, which literally means, Place Gathering Driftwood.[33]

KE ALA I KAHIKI

The Road to and from the Horizon

Kealaikahiki channel from Heiau Ka'iwaloa with Lāna'i on the right

On the western tip of the island now called Kaho'olawe, Hawai'i Loa landed his canoes on a sandy beach and named the promontory and its sea channel Ke-ala-i-kahiki, The Road-to-and-from-the-Horizon (or Tahiti).

As they traveled up the channel they saw a small barren island and named it Lā-na'i, or Conquered-by-the-Sun-and-Tides. Then they saw a large fertile island before them, crowned with tall mountains shrouded in misty clouds that seemed to be like the Juncture between Heaven and Earth, so Hawai'i Loa named the mountain range Mauna Kahālāwai and the island for his son, Mā'ui, Gathering Beauty.[34]

Mauna Kahālāwai

ʻUKUMEHAME VALLEY

As they steered their canoes toward land, the valley before them was lush with streams and plant life. They landed their canoes and gave thanks for safe passage to such a place of beauty and abundance. They named the valley ʻUku-me-hame, in recognition of the forces that brought them safely over the vast ocean; *ʻUku* (reward) *me* (from) *hame* (Creator's four corners or pillars of the sky). In honor of the guide stars that brought them to this distant land, they named the peak to the south side of ʻUkumehame Valley as Puʻu Hōkūʻula, or Red Star Hill, for Hōkūʻula, or Aldebaran, the red star "eye" of Taurus, and the peak in the heart of the valley they named Hōkūʻula iki (small red star) for Mars.[35]

When the canoes of Hawaiʻi-loa's brother Kiʻi joined them, they were overjoyed. A great storm had separated them at sea. They built twin *heiau* (temples) of praise and thanksgiving on each side of the valley entrance and named the twin *heiau* as He-kiʻi, meaning a tryst or meeting.[36]

From Heiau Hekiʻi towards Kahoʻolawe

OLOWALU VALLEY

From Heiau Ka'iwaloa looking up to Olowalu Valley

At the base of the neighboring valley which Hawai'i Loa called Olo-alu (Working Together), and now called Olowalu (Many Hills), stands the Heiau Ka-'iwaloa, overlooking the navigational channel, Kealaikahiki, between Kaho'olawe and Lāna'i. This is a Hale Kilo Hōkū, or star observatory, attributed to Hawai'i Loa and his people. It is said to be named for the *'Iwa,* or frigate bird, that helped guide them to this land. Heiau Ka'iwaloa was a temple of astronomy and astrology and was also used for healing because of the warm water springs nearby.

Heiau Ka'iwaloa overlooking Kealaikahiki channel with Lāna'i in the background

HEIAU KAʻIWALOA

Heiau Kaʻiwaloa at the mouth of Olowalu Valley

Kilo hōkū (astronomers) studied the stars here and taught astronomy and navigation. These astronomers were *kāhuna* (priests) of the *aliʻi* class and possessed great *mana*, or spiritual power, in their knowledge of the stars, the winds, tides, currents, fishing, and health. The warm water spring of Olowalu, named Wai-wela-wela, (*wai* – water; *wela* – hot) was used for healing.[37]

At this temple the *kāhuna kilo hōkū* prayed to the demi-god Kuhi-mana. The *kāhuna lapa-au* (medical priests) prayed to the diety Maiola (*mai* – come; *ola* – life) and his female companions Kau-ka-ho-ʻō-la-maʻi and Kapu-ala-kai.[38]

Navigator on the Observatory.
Painting by Herb Kane

PUʻU KĪLEʻA

Puʻu Kīleʻa overlooking Kealaikahiki Channel. with Lanaʻi in background

Below Heiau Kaʻiwaloa is a cinder cone Hawaiʻi Loa named Puʻu Kī-leʻa. Lea is a navigational star in the constellation Pleiades, named for Makaliʻi (Chief's Eyes), but is also the name of the goddess Lea, revered by canoe builders. The goddess Lea could take the form of the *ele-pai-o* bird to show canoe builders which *koa* trees were best for canoes. Said as *leʻa*, the word means to observe clearly, without error. *Kī* means to aim.

THE LEGEND OF HĀLOA

One legend told about Puʻu Kīleʻa involves Hāloa, the first-born son of Wākea and Papa-Haumea. Also known as Hāloa-naka, the infant died at birth and was buried atop Puʻu Kīleʻa. Soon a *kalo* (taro) plant sprouted from the child's body. The leaf was named Lau-ka-pa-lili, the Quivering Leaf. The stalk was named Hāloa for the infant child.[39] Every island has a story about Hāloa. Many Hawaiians still consider the *kalo* plant as a sacred genealogical ancestor.

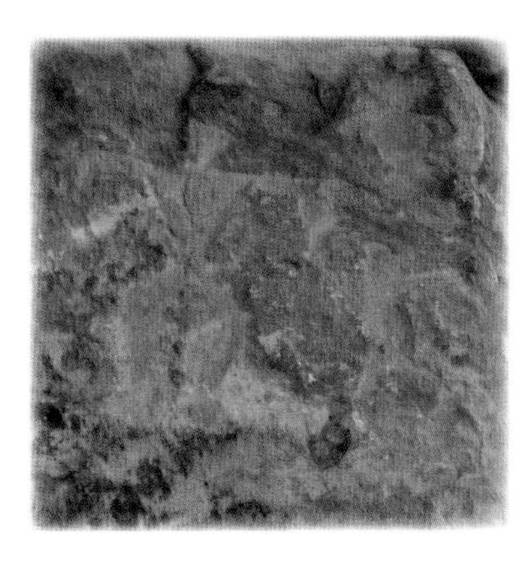

Olowalu Petroglyph

The Lahaina side of Pu'u Kīle'a

The Lahaina side of Pu'u Kī-le'a is exposed basaltic rock covered in petroglyphs (figures etched in stone), many of which are attributed to Hawai'i Loa's first expedition. They include a sailing canoe, a bird-headed figure, a circle, and a circle with a dot. Because the bird-headed man is similar to the Egyptian god Horas, some people in the civic club believed that Hawai'i Loa came from the area of the Nile.

Located adjacent to the trailhead into Olowalu Valley and nearby freshwater springs, the basaltic rock is covered, undoubtedly, in centuries of images.

Eventually Hawai'i Loa would come to discover and name all eight islands and eight sea channels, naming the biggest island Hawai'i, for himself and the rest after his children.[40]

Ki'i soon returned to the south and peopled Tahiti, Ra'iatea and Mo'orea. Hawai'i Loa made many return trips to the south and brought back the offspring of Ki'i to mate with his children. It is said that the descendants of Ki'i, Hawai'i-Loa, and Makali'i peopled the entire Hawai'i group. From Hawai'i-Loa descend the *ali'i* chiefs and *kāhuna* class and from Makali'i descend the *maka'āinana* (commoners).[41]

The story of Hawai'i Loa is found in the Kumuhonua Genealogy with various regional versions told on Hawai'i and Maui. These stories were handed down orally through generations of *ali'i* and their descendants on Moku Maui. Mrs. Ashdown first heard the story as a child from Kahu Kīna'u at 'Ulupalakua Ranch and then from various members of the West Maui Hawaiian Civic Club. The fact remains, someone gave the place names long ago and the names endure like monuments to the people of antiquity who found this place in the middle of the vast Pacific.

MAUNA LĪHAU

Mauna Līhau stands to the left of Olowalu Valley

THE LEGEND OF LĪHAU

Long ago, when demi-gods walked the earth, the *ali'i wahine* (woman) named Lī-hau lived in Kahiki with her husband Ke'eke'ehie. They had a very beautiful daughter named Koa'e-loa who was being pursued by four chiefs of inferior rank who wanted to steal her away. Līhau came with her husband and daughter to Maui to escape the wicked chiefs, but they followed, giving Līhau and her family no peace. When the noted priest of Lele (now Lahaina), Wa-o-lani, heard of their plight, he changed the tormented family into mountain peaks. Līhau is the mountain on the left side of Olowalu Valley and Ke'eke'ehie is the ridge above. The daughter, Koa'e-loa, is next to her father above Kaua'ula Valley. Kahuna Wa-o-lani turned the suitors into cinder cone hills at their feet, Pu'u Hipa, Pu'u Kahea, Pu'u La'ina, and Pu'u Keka'a.[42]

There are many stories told of Līhau in *mele* (song) and chant. This story was told by William E. Saffery,[43] a descendent of Sea Captain Edmond Saffery (c. 1839) and Kawa'a'iki Alapa'i Na'ehu, the granddaughter of King Alapa'i-nui. Saffery lived on royal land near the Loko o Kapa-iki, Queen Kalola's fishpond in Olowalu. Queen Kalola's lands were entrusted to the *konohiki* (Chief caretaker) *ali'i* Na'ehu after her death.[44]

Aliʻi-wahine Līhau

Līhau can be seen on the mountain in the moonlight as a beautiful woman, lying on her back with her hair streaming down to Olowalu Valley.[45] Kapela is the highest peak of Mauna Līhau.

PŌHAKU MANU

Pōhaku Manu on Pali Kila

On Mauna Līhau is a cliff outcropping overlooking Olowalu Valley named Pali Kila, meaning symbolic cliff or emblem. On this cliff is a stone formation that appears like a hovering bird and is named Pōhaku Manu, or Bird Rock. Some say it is the *ʻIwa,* the seabird that guided the explorers to land during the last days of their journey. Hawaiian Civic Club members said there were numerous petroglyphs on Pali Kila. When the sun or moon is right, you can easily see Pōhaku Manu up on Mauna Līhau, hovering over Olowalu Valley and the Heiau Kaʻiwaloa.[46]

PUʻU KAPŪʻALI

Below Mauna Līhau is a rock promontary jutting into the dual waters of Awalu (Twin channels or harbors). Named as Puʻu Ka-pū-ʻali in the 18th century to commemorate the place where the forces of Kamehameha-nui of Maui and his uncle, King Alapaʻinui of Hawaiʻi, joined together to prepare for battle with Kamehameha-nui's half-brother, Kau-hiʻai moku-a Kama, who was challenging him for the Maui realm.

Kapūʻali refers to a warrior whose *malo* (loin cloth) is tied for warfare, with no loose ends, and is remembered as the place where the bloodiest war in Maui's history began, brother against brother. Literal translation of Puʻu Kapūʻali is Hill of the Warrior.

In the early 1700s, the Ke-kau-like Dynasty, ancestors of the Piʻilani Dynasty, ruled all of Maui, Molokaʻi, Lānaʻi, and Oʻahu, and eventually Kauaʻi through marriage. Kekaulike had many consorts of rank and produced 15 children during his life. His firstborn son, Kauhiʻaimoku a Kama (*ʻaimoku* – heir and ruler, *a* – of, *Kama* – Maui) was Kekaulike's noted commander-in-chief who aided his father in establishing peace by winning many battles for him and so expected to inherit from his father.

However, Kekaulike's highest ranking wife was Ke-kuʻi-apo-iwa-nui, his own half-sister *and* the sister of King Alapaʻinui of Hawaiʻi. With this divine chiefess, King Kekaulike had three *nīʻau piʻo* (divine) children; Kamehameha-nui, Kahekili II, and the Princess Kalola.

A Canoe of the Sandwich Islands.
John Webber. From *1785 Voyage of the Pacific Islands,* G. Nichol and T. Cadell, London, 1785

King Alapaʻinui had been trying for decades to unseat his brother-in-law King Kekaulike and take Maui's considerable realm. In 1736, as Kekaulike lay dying in Hāna, he knew Alapaʻinui was preparing to invade once again and so in hopes of avoiding war, he named Alapaʻinui's favorite nephew, and his higher ranking son, Kamehameha-nui, as heir. This averted any warfare between Maui and Hawaiʻi as Kekaulike had hoped, but Kauhi was resentful and in 1738 he rallied support from Chief Pele-io-holani of Oahu to challenge Kamehameha-nui for the Maui realm.

Kamehameha-nui and Alapaʻinui met at Puʻu Kapūʻali and refreshed their troops at the spring there called Maka-lua-puna before advancing forward to the Battle of Kōkō-o-nā-moku (Blood of the Islands) at what is now Kaʻanāpali.[47]

Once thought to be a resting place of Madam Pele, until her sister Kaipo created the fresh water spring of Makaluapuna, the ancient name of Kapu-ale was used in reference to offerings and prayers to the dieties who once inhabited this place.

HEKILI POINT, OLOWALU

Lae Hekili

Lae He-kili (Thunder Point) below Olowalu Stream was named for King Kahe-kili-nui 'Ahumanu II, Kamehameha-nui's brother, and the last King of Maui.

The beach from Lae Hekili to Pahoa stream at Puamana Park is named Ke-awa-lua, (the two channels or harbors). The reef along this shoreline is broad and the sea is shallow and the reefs act as a natural barrier to the shore, which early Maui defenders used to their advantage, burying *pahoa* (wooden daggers) in the coral to puncture advancing canoes. There were only two openings in the reef where canoes could pass without fouling on the coral, hence the name, Keawalua.[48]

Beyond Lae Hekili is the remnant of Olowalu wharf, once supporting the shipping needs of the Olowalu Mill and sugar plantation in the late 1800s. In 1931 the entire mill was dismantled and shipped to the Phillipines.

Mrs. Keaumoana of Olowalu told Mrs. Ashdown that the name of the old wharf was Ka-lua-aha.[49]

Olowalu Wharf. Photo by Forest and Kim Starr

LAUNIUPOKO VALLEY

Photo by Forest and Kim Starr

As Kamehameha-nui and his uncle, Alapa'inui, rallied their troops at Pu'u Kapū'ali, Kauhi'aimoku-a-kama and his warriors gathered in Lau-niu-poko with the troops of O'ahu's new king, Pele-i'o-ho-lani. From here, Kauhi sent his brother a coconut frond as a challenge to war and Kamehameha-nui returned the palm broken off short as a sign that he accepted the challenge. Lau-niu-poko (short coconut leaf) commemorates Kauhi's brave challenge for the Maui realm and Kamehameha-nui's answer.[50]

In Launiupoko Valley was a *pākau'a*, or fortress, a round stone tower-like structure. Kauhi and Pelei'oholani held off Kamehameha-nui and Alapainui's forces in the first siege by damming the waters of the valley, causing a shortage of water for the enemy while they retreated to the fortress and refreshed their troops with *wai-niu* (coconut water) and the meat of the coconut. Kamehemeha-nui's warriors were forced to leave. In this instance, Launiupoko is also in reference to Kauhi's strategy in holding off the opposing warriors.[51]

PŌHAKU PEPEIAO MOʻO

On the southern end of Launiupoko Beach Park is a large stone protruding from the nearshore waters named Pōhaku Pepeiao-moʻo, and generally translated as Lizard's Ear (*Pepeiao* – ear, *moʻo* – lizard). Aunty Kai took Mrs. Ashdown here to explain the true meaning, based on the *hūnā* translation. "Today, people do not remember the secret meanings of our language. *Moʻo* also means genealogy and *Pepe-au* pertains to the sea-currents and the ancestry of the ocean. The ancients taught that all life came, originally, from the sea." A slight misspelling, while sounding the same, can result in a completely different meaning. The true meaning of the name is Pepe-au-moʻo (Ancestral-Spirits-of-the-Ocean) and the stone was once a beautiful maiden called Ua-ko-ko, named for the low-lying red rainbow seen in the sea mists. Her sea-spirit ancestors turned her into a stone to protect her from her lover's jealous mother. Now the Red Rainbow of the Sea Mists lies protected by the sea dieties where they surround her with the fragrant sea-moss called *limu-līpoa*.[52]

KAUAʻULA VALLEY

Sacred Battle Valley

As Kamehameha-nui's forces advanced by land and sea towards Lahaina, Kauhi retreated to the fortress Kā-hili at the base of Pali Kāhili (Royal Standard Cliff) in Kauaʻula Valley (Sacred Battle Valley). From there he deployed his forces to meet Kamehameha-nui and Alapaʻi-nui at Keʻawaʻawa (now Kaʻanāpali) where the Battle of Koko-o-nā-moku (Blood of the Islands) left many *aliʻi* dead, including Kauhiʻaimoku-a-kama.

Kamehameha-nui's rule over Maui was restored and he maintained peace in the Maui Kingdom for 29 years until his death in 1765. Kamehameha-nui-ʻAi-luau left the Maui Kingdom to his brother, Kahekili-nui-ahu-manu II, who ruled until his death in 1794.[53]

The Kauaʻula Stream provided early irrigation for the many fishponds and *loʻi kalo* (taro patches) of Lahaina, as well as the sacred pond of Moku-hi-nia in the heart of Luaʻehu, home of the *aliʻi*. In the early 1400s King Piʻilani had a stone paved *ʻauwai* (irrigation ditch) constructed that ran from Kauaʻula Stream to above Lahaina. Known as ʻAuwai o Piʻilani, its waterways irrigated lands on both sides of the Kauaʻula Stream.[54]

THE KAUAʻULA WIND

Rain Falling West Maui Mountains, Kauaʻula Wind, Luaʻehu.
Photo by Ray Baker. R. J. Baker Collection, Bishop Museum

From Kauaʻula Valley blows a wind made famous in *mele* (songs) for its power of destruction. It was known to blow with hurricane force in the 1800s and has been clocked at over 120 miles per hour. Coming from Iao Valley over Pali Kāhili, down Kauaʻula Valley, the Makani Kauaʻula (Sacred Battle Wind) could blow from Honokowai to Waikapu, tearing down everything in its path. In 1858 it famously tore off the stone steeple and half the roof of the Waineʻe Church. A hundred years later, in 1951, it completely demolished the second church, built in 1894 by H. P. Baldwin to replace the original structure burned out by anti-Royalists in 1893.

Mrs. Kaʻae (Aunty Kai) reported that the people were given a warning when the hurricane winds would blow. A whistling sound would be heard blowing from the valley three to five days before heavy winds began and people would have time to tie down their houses and evacuate with their belongings. Many sought shelter in big caves. One such cave was in Wahi-kuli Valley, where the floor was paved with smooth black pebbles.[55] Many chants tell of the destructive powers of the Kauaʻula Wind, how "without teeth, the wind is destroying food from Pualo to Moaliʻi."[56]

MĀKILA POINT

Kaua'ula Stream reaches the ocean at Mākila Point, now known as Puamana

Mā-kila beach runs from Mākila Point to Kamehameha III School in Lahaina. The Heiau Mākila once stood where the Puamana complex now stands. Mākila was said to relate to "very high places" and this temple complex was a place of inspiration for various *kāhuna*. It was somehow connected to Heiau Ka'iwaloa, the temple of astronomy, and the strange stone towers in Launiupoko and Kaua'ula Valleys.[57]

HEIAU WA'ILEHUA

On this point, perhaps on the foundation of Heiau Mākila, it is said the Heiau Wa'i-lehua (Broken Warrior) (*wa'i* – to break, *lehua* – warrior) was built by Kama-lala-walu (c. 1575) as a *luakini* (sacrificial) heiau. Next to the heiau were altars where victims of sacrifice were slain.

It was here that Kauhi'aimoku-a-Kama made the fateful decision to challenge his brother Kamehameha-nui for the Maui realm. Kauhi was carrying stones to reconstruct Heiau Wa'ilehu when the ambitious *kāhuna* (priest) named Pi-na'au told him; "Let the weak carry the stones. The work of the strong is to establish themselves upon the land." The *kāhuna* told him, "Go to war, stand at the head of the government." Unfortunately, Kauhi followed his advice and was slain in the Battle of Koko'onāmoku. Afterwards, Kamehameha-nui honored his half-brother Kauhi'aimoku-a-Kama with burial rites at Heiau Wa'i-lehua, where his body was burned.

On the moon nights of Kāne and Lono many people in contemporary times have heard the sounds of drums and chanting here.[58]

OLOWALU MASSACRE

Lahaina 1854 .
Robert Elwes. From *A Sketchers Tour Round The World,* Hurst and Blackett, London

After Cook's arrival in the islands in 1778, fur traders from the northwest coast of America began to stop in the Islands for provisioning on their way to China. In 1790, the American ship *Eleanor*, under the command of Captain Simon Metcalf, anchored off the coast of Honua'ula, on his way to rendezvous with his sister ship, the *Fair American*, captained by his son Thomas, at Kona, Hawai'i.

Living in Honua'ula (on the western slope of Haleakala) at the time was dowager Queen Kalola, daughter of King Kekaulike and sister to his heirs, Kamehameha-nui, and Kahekili, The Thunderer, who was now ruling from O'ahu. High

Chiefess Kalola had been married to the High Chief Kalaniōpu'u, brother of Alapa'inui, the Ruling Chief of Hawai'i. After the death of Kalaniōpu'u, Queen Kalola came home to Maui and took Chief Ka'ōpuiki as her husband.

Trading with the sailors went well, with Kalola and Ka'ōpuiki in attendance, until a crewman slashed at Chief Ka'ōpuiki with ropes as he tried to clear the ship. Ka'ōpuiki sought revenge for the insult and that night he and his men killed the watchman and stole the ship's long boat.

In the morning Captain Metcalf sailed up the coast, believing that the culprits had been hog traders from Olowalu village. Queen Kalola was the highest ranking chiefess on Maui and was the only *ali'i* who carried the *Kapu Mau'u-mae* (withered grass), implying destruction, and had the penalty of death by burning for anyone who broke her *kapu*.

Kalola sent her runners to Olowalu and declared a *kapu* on trading to keep the people away from Metcalf's ship in order to avoid bloodshed. After a few days the *kapu* was lifted and natives approached the ship to resume trading. Metcalf directed all their canoes to one side of the ship. He then ordered his gun deck ports opened and gave the order to fire. At least 100 natives were massacred and more than a hundred more were wounded. Metcalf weighed anchor and sailed for the island of Hawai'i to meet with the *Fair American*.

Word of the massacre reached Hawai'i ahead of the *Eleanor* and the High Chief Ka-me'e-ia-moku organized an assault on the small ship anchored off the west coast of Hawai'i. Sneaking aboard the ship at night, they killed the crew, except for one man, Issac Davis, and sunk the *Fair American*. Davis was taken to Kamehameha I in Kona and secreted away and a *kapu* was placed on the people to keep Metcalf from discovering what had happened. When he arrived and found no natives to welcome him, Metcalf sent John Young ashore to seek information on the *Fair American*. Young was captured and when he did not return to the ship, Metcalf sailed away without Young or his son.

The *Fair American* was raised and refitted for Kamehameha's use as a war ship. Davis and Young became advisors (and gunners) to Kamehameha I, who began amassing guns and, along with the sloop *Fair American*, were instrumental in his success in conquering Maui and uniting the islands under one peaceful kingdom.

Both Young and Davis married high ranking relatives of Kamehameha I and were made chiefs who served in His Majesty's government the rest of their lives, leaving families whose ancestors still live on Maui. John Young's granddaughter became Queen Emma, wife of Kamehameha IV.[59]

THE BATTLE OF KAPANIWAI O ʻĪAO

Kamehameha commands the Fair American. Painting by Herb Kane

In July of 1790, five months after the Olowalu Massacre, Kamehameha brought all his forces against Maui in the decisive battle of *Ka-pani-wai o ʻĪao* (The damming of the waters of ʻĪao). Kamehameha's war machine of twenty thousand warriors included John Young and Issac Davis at the helm of the *Fair American,* deploying the cannons and artillery. The Maui troops, led by Kahekili's son, Ka-lani-kū-pu-le, were driven into ʻĪao Valley where the waters were dammed with the bodies of slain warriors.[60]

Observing the battle from a plateau high in ʻĪao Valley was the dowager Queen Kalola, with her two daughters, and her eleven-year old granddaughter, Princess Keʻōpūolani, the highest ranking of all *aliʻi* at the time. Born of the daughter and son of Kalolo, this *nīʻaupiʻo aliʻi* was pure blood of both the Piʻilani Dynasty of Maui and the Umi Dynasty of Hawaiʻi, giving her divine status.[61]

Accompanied by Prince Kalanikūpule and his chiefly retainers, Kalola's party escaped to Lahaina over Pali Kāhili (Cliff of the Royal Standard), or perhaps a secret lava tube through the cliff, resting at Puʻu Paʻupaʻu, the place of refuge above Lahaina. The ailing Kalola and her party went by canoe to Kalamaʻula on Molokaʻi, the sacred place named for their family *ʻaumakua,* Kihawahine Mokuhinia Kalamaʻula. Prince Kalanikūpule joined his father King Kahekili on Oʻahu.

PALI KĀHILI

Kamehameha I wanted the highest ranking divine *ali'i* in all Hawai'i, Princess Ke'ōpūolani, to be the mother of his heirs, securing their high born rank, and therefore the *mana* (spiritual power) of the future Kamehameha Dynasty. He sent messengers to Moloka'i to offer protection to Queen Kalola and her family in exchange for the sacred Ke'ōpūolani. Kalola agreed and the young princess, along with her mother and aunt, were given to Kamehameha at Waihe'e where the truce, or Kiss of Peace took place. Named Ke Honi (The Kiss), the area is now a municipal golf course.[62, 63]

Kamehameha I sent a message to Kahekili II on O'ahu in the form of two stones, one white, one black. Kahekili II responded to the messengers, "Tell Kamehameha to return to Hawai'i and when the black *kapa* (cloth) covers Kahekili and the black pig rests at his nose, this is the time to cast stones." Spoken as a parable in the *hūnā* dialect of the *ali'i*, the message was clear; "When I die, the kingdom will be yours."[64]

Four years later, in 1794, the Maui kingdom's final ruler, Kahekili-nui 'Ahumanu, The Thunderer, died at the age of eighty-seven on O'ahu. His body was claimed by his half-brothers, the twin guardians of Kamehameha I, High Chiefs Ka-me'e-ia-moku and Kamanawa.

After the death of Kahekili, Maui, Moloka'i, and Lāna'i were reconquered by Kamehameha I in 1794. In 1795 he took O'ahu, ruled by Maui kings since 1780, by defeating the forces of Kahekili's son, Kalanikūpule.[65] With sacred Maui Chiefess Ke'ōpūolani, Kamehameha I would begin the Kamehameha Dynasty that would last for a hundred years.

LAHAʻĀINA

Lahaʻāina and the ʻAuʻau channel

"They said we must be careful to pronounce correctly, the old way, and repeated several times 'Lahaʻāina, Land of Prophecy'." (Queen Liliʻuokalani and party to Inez and her mother upon entering Ke-awa-iki harbor aboard the steamship *S.S. Mauna Kea* in January, 1908.)[66]

Laha means prophecy, or something published, like a law or a proclamation. *ʻĀina* means land; Land of Prophecy. Hence the name La-ha-ai-na, more properly spelled and pronounced as Lahaʻāina and said almost as two words. This pronunciation was important to Queen Liliʻuokalani as she had Inez and her mother repeat it several times while she introduced them to Lahaʻāina in January 1908.

Inez and her mother traveled to Maui aboard the *S.S. Mauna Kea* with the Queen and Her Majesty's party, who were disembarking at Lahaʻāina while the Ashdowns went on to Mākena. As they entered the ʻAuʻau channel approaching Lahaʻāina, one elderly aristocratic lady tossed her head and said, "Lahaina Roads, indeed! ʻAuʻau means our entire background since the beginning of creation. How will our children remember the beauty of our island culture when even our melodious Hawaiʻi names are forgotten?" Whalers had long before renamed the channel Lahaina Roads, or Roadstead.

Queen Liliʻuokalani replied, "Some will remember."[67]

As the Queen disembarked, little Inez watched as the crowd ashore fell to their knees and began wailing a greeting to their Queen in the ancient fashion. Some reached out and kissed the hem of her dress as she slowly walked to her carriage. It must have made quite an impression on the eight year-old, because she decided then she would remember the "old names."

Modern pronunciation has become "La-hai-na," said as one word, Lahaina, rather than Laha'āina. Lahaina means "cruel sun" and comes from an old wive's tale about a chief in the area who declared "what a cruel sun!" Lahaina is easier to pronounce than Laha'āina, so eventually ease prevailed.[68] But the elders of the West Maui Hawaiian Civic Club told Mrs. Ashdown if a name is mispronounced this way, then it is like "calling out" to have a cruel sun. This was the land of prophecy, where even the first written laws of the Kingdom were proclaimed.[69]

Before the great prophet Laha'āina-loa gave his name to Laha'āina, the area near Ke-awa-iki (the small harbor) was known as Lele, the favorite place of "coming and going" where canoes of *ali'i* and commoners alike came to enjoy the surf of 'Uo, even more favored than the surf of Ka-lehua-wehe at Waikīkī.[70] Centrally located between Hawai'i and O'ahu, the village became known as Lele, the place of Coming and Going. The famous surf of 'Uo was disrupted by the Lahaina Harbor and breakwater, built in the early 1950s, but the surf of 'Uo can still be seen along Mākila beach.

The 'Au'au (Ancestral) Sea channel, between Laha'āina and Lāna'i, is the Ancestral Sea of deep currents and refers to the Era of Creation. The Twin Currents of Hawai'i's eight sea channels rise in 'Au'au and travel among the islands, returning here according to the moon and tides, bringing the Ma'a'a breeze that blows from Ma'alaea Bay.[71]

The surf of Uo, "Surf board, a bathing scene, Lahaina," 1855.
James Gay Sawkins. 2492605. National Library of Australia.

Ke-au-ka is the outgoing current coming from the shore and Ke-au-miki is the incoming current going to the shore. Ancient canoe lanes followed the Twin Currents and the schools of fish that traveled with Keauka and Keaumiki. It is said that the Twin Currents were named for the twin sons of the noted teacher, Kū-a-moana-hā, (Kū of the four oceans) who taught the secrets of the sea from ʻĀina o Kanaloa, now Kahoʻolawe. The ʻAuʻau channel merges at Kahoʻolawe with the Kealaikahiki sea channel, the Road to and from the Horizon.[72]

In the time of the gods, Kapō (or Kaipo), goddess of the Deep Sea, daughter of Haumea Papa and Wākea, brought the *ʻulu* (breadfruit) tree to Lele where it flourished and provided wonderful shade and nutritious fruit throughout the village. Soon, Lele became known as Malu-ʻUlu-o-Lele, "Lele – in the Protective Shade of the ʻUlu."[73]

When the missionaries arrived in Lahaina in 1823, the village was still noted for its breadfruit trees, as the Reverend C. S. Stewart described in his Journal; "The thick shade of the breadfruit trees surround our hosts' cottages—the rustling of the breeze through the bananas and sugar-cane—the murmurs of the mountain streams encircling the yard—and the coolness and verdure of every spot around us—seemed like the delights of an Eden."[74]

Large ʻUlu tree, Dance of the Women of the Isles of Sandwich.
Louis Choris. From *Voyage pittoresque autour de monde,* F. Didot, Paris, 1822

LUAʻEHU

Mākila Beach at Luaʻehu

Luaʻehu was the restricted area in Lahaina where only the *aliʻi* lived. Fronting Mākila beach with its famous ʻUo surf, the lands exclusive to the *aliʻi* went north to Pahu-mana-mana stream, now Dickenson Street, and south to Pahoa stream, now Puamana Park, and into the uplands. The name refers to the colorful, splendid ones, the *aliʻi*, who made Luaʻehu a favorite home for centuries.[75] In the *hūnā* dialect, it refers to the ancestral *ʻaumakua* Kiha Wahine Mokuhinia, who lived in the pond named for her in Luaʻehu.[76]

All along the shoreline of Luaʻehu the large *hale pili* (grass houses) of the *aliʻi* were shaded by *ʻulu* (breadfruit) trees, *kou* trees with red-gold blossoms, and hundreds of coconut palms. This was the meeting place of the chiefs of the Maui Kingdom and home to many of the ruling *aliʻi*.[77] It was here that Kamehameha established his first Capital of the Kingdom of Hawaiʻi.

Where Kamehameha III School and Holy Innocents Church now stand along the Mākila shoreline, were the homes of the *aliʻi*. Queen Liliʻuokalani, Hawaiʻi's last monarch, lived here as a child with her adopted parents Pākī and Kōnia, and her adopted sister Bernice Pauahi Pākī, who would later leave the large Bernice Pauahi Bishop Estate for the benefit of the Hawaiian people[78]

LOKO O MOKUHINIA

Photo by Ray Baker. R. J. Baker Collection, Bishop Museum

The entire Lua'ehu area had once been wetlands in the flood plain of Mauna Kahālāwai. The many streams and rivers were channeled into fishponds and *lo'i* (taro patches). Flowing from Kaua'ula Valley, the Mokuhinia River fed a large pond called Mokuhinia (now Malu-o-lele Park) and then the royal fishpond called Wai-o-kama before reaching Mākila beach. Named for King Kama-lala-walu (c. 1575) for whom Maui became known as Maui-a-Kama (Maui of Kama), this fishpond was one of many *loko i'a* (fishponds) bordering Mokuhinia pond.[79]

King Kama-lala-walu was the son and heir of Kiha-Pi'ilani and a contemporary of Chief Lono-i-ka-makihiki of Hawai'i. Lono sent "deserters" to Maui to falsely claim his weakness in order to lure Kama-lala-walu into battle. Kama was overwhelmed and betrayed by trusted warriors. He fought bravely and died at Puako, Hawai'i. His son, Ka-uhi-a-Kama, escaped. To honor the reign of this brave king, Maui became known for a time as Maui-a-Kama (Maui of Kama).[80]

Mokuhinia River eventually became known as Kaua'ula Stream.

MOKUʻULA

Mokuʻula Island, with Waineʻe Church, "The Presbyterian Church, Lahaina," 1851.
James Gay Sawkins. 2492653. National Library of Australia.

Loko o Mokuhinia (the pond of Mokuhina) was at the center of Luaʻehu and was *kapu* (forbidden) to all but the ruling *aliʻi*. In ancient times the pond extended southward much further than present Shaw Street, and to the town-side past Chapel Street to Pahumanamana Stream (now Dickenson Street), and contained many islets on which houses of the *aliʻi* were built. Some parts of the pond were very deep. The largest islet was an acre in size and named Mokuʻula (Sacred Island). On it was a large *hale pili* (grass house) and an eating and meeting *hale* for the exclusive use of the ruling *aliʻi*.[81]

Since the Mokuhinia pond was named for the Kiha-Wahine Mokuhinia, who in life had been a daughter of King Piʻilani (c. 1525), it seems likely that Mokuʻula would have been inhabited since at least the reign of King Piʻilani.[82]

Mokuʻula (sacred island) was guarded by the personal *ʻaumakua* of the Maui royal family, the *moʻo* goddesses of the Kiha.

KIHA-WAHINE

Statue (ki'i) *of Kiha-wahine.*
Drawing by Robert C. Barnfield, 1855.
Bishop Museum

Known as the *mo'o* in the *Mo'o-kū-'auhau* (Genealogical lines of *ali'i*), these *'aumakua* were ancestral spirits transformed at death into *mo'o* (lizards or dragons) with supernatural powers. They lived in fresh water springs and streams and ponds and assisted their mortal relatives.[83]

Most always female, these *mo'o* goddesses can be found in the oldest tales throughout Hawai'i and Polynesia. Like the interlocking vertebrae of the lizard, *mo'o* also refers to genealogical succession found in the Mo'okū'auhau.[84]

The *ali'i* families of Maui had *'aumakua* from the famed Kiha Family of *mo'o* goddesses, originating in the Ulu-Hema genealogy with Mo'o-i-nanea.[85] They often appeared in visible form on certain occasions and when wrongs were contemplated or had been done. They warned their descendants of coming disasters and when a descendant was about to die. As a group, the Kiha dieties of Maui were referred to as the *Kia'i*.[86] *Kia'i* means to watch, to guard.

There are many stories told of the deeds of the Kiha Family in *mele* (song) and *oli* (chants) across Hawai'i. Among the most noted Kiha of Maui were Mo'oinanea, the famed ancestress of Luaʻehu known as the "Self-reliant Dragon," Wao, who lived in the *ʻauwai* (irrigation ditch) named Wai-o-Wao by Kanaha Valley, and Kiha-wahine Mokuhina, the last *moʻo* goddess of the Kiha Family, said to be seen by hundreds at her home in Mokuhinia Pond in 1838.[87]

In life, Kiha-wahine Mokuhinia was the *nīʻau piʻo* (divine) daughter of King Piʻilani of Maui and lived in the mid-16th century.[88] After her death, Kiha-wahine Mokuhinia was deified in an elaborate ritual called *kā-kū-ʻai*. Her *iwi* (bones) were wrapped in yellow *tapa* (bark cloth) dyed with *ʻōlena* (turmeric) and ceremoniously layed into the waters of Mokuhinia. At this moment, her spirit transmigrated into a *kiha* goddess whose chosen name became Ka-lama-ʻula (Torch of the Heights). Only the sacred wood of the lama tree was to be used for her altars.[89]

After deification, Kiha-wahine Kalamaʻula made her home in a cave called Ka-lua-o-kiha (The Pit of the Kiha) below the waters of the pond named for her, Mokuhinia, where she protected her *aliʻi* ancestors who lived on the sacred island, Mokuʻula. Depending on the genealogy, she was called Kalamaʻula Mokuhinia, Ka-la-maʻi-nuʻu, or Ka-lai-ma-nuʻu.[90]

Through a succession of *piʻo* (brother-sister) marriages of Piʻilani descendents, the noble *nīʻau piʻo* Princess Keʻōpūolani was born of the highest bloodlines of Maui and Hawaiʻi. Considered the highest ranking *aliʻi* of the time, Keʻōpūolani was also the last ancestress of Kiha-wahine and so possessed the *mana* (spiritual power) of her personal *ʻaumakua,* as well as the *mana* of her own divine rank.[91]

When Kamehameha I conquered Maui in 1790, he took Keʻōpūolani as his prize, and through this alliance he also won the protection of her *ʻaumakua*, Kiha-wahine.[92] Kamehameha built altars to Kiha-wahine and had her carved image carried during the Makahiki procession of Hoʻoilo, the Wet Season. Kiha-wahine was the only female deity ever afforded such a privilege.[93]

Each generation of *aliʻi* had a *kahu* (guardian) for the *kiha* goddess. Governor Hoapili (Ulu-mā-hei-hei) was the final *kahu* for Keʻōpūolani's *ʻaumakua* Kiha-wahine Kalamaʻula Mokuhinia. After Kamehameha's death in 1819, at his request, Governor Hoapili married Keʻōpūolani and became the *kahu* to her *ʻaumakua* Kiha-wahine.[94]

Before her death and deification, Chiefess Kiha-wahine Mokuhinia was married to Kamalama and they had Nihoa Kamalama. Nihoa had Maluna, from whom seven generations later descended Hawaiʻi's final rulers, King Kalākaua and Queen Liliʻuokalani, all related by blood to Kiha-wahine of the Piʻilani Dynasty.[95]

Aunty Kai was witness to the disinterment of the *aliʻi* in the stone mausoleum on Mokuʻula in 1884. Princess Bernice Pauahi Bishop had ordered their remains removed to Waineʻe Cemetery. Aunty Kai, and her sister Kawaiʻele, remembered watching the *aliʻi* in their colorful feather capes paddling their canoes on Mokuhinia Pond to Waineʻe Church as late as 1893.[96]

Mokuhinia, 1910. Photo by Ray Baker. R. J. Baker Collection, Bishop Museum

Beginning in the mid-1860s, construction of ditches and reservoirs above Waine'e for sugar irrigation began disrupting the flow of waters through Mokuhinia and the edges of the pond began slowly filling with sedges and reeds. By 1897 the pond became stagnate and a breeding area for mosquitoes.[97]

About 1913, Mr. Weinseimer, manager of the Olowalu and Pioneer Plantations and George Freeland, owner of the Pioneer Hotel, urged government officials to improve Lahaina. With the help of plantation laborers and equipment, the Mokuhinia Pond and its sacred island were unceremoniously filled in and the area made into a park named Malu-'ulu-o-lele, in honor of the old name for Lele.[98] Only the remnants of Wai-o-kama fishpond, fed by the waters of Mokuhinia, remained of the home of Kiha-wahine.

Alice Ka'ehu-kai Shaw Ka'ae (Aunty Kai) and her sister, Mary Kawai'ele Shaw Hoapili had inherited the southern sections of Wai o Kama from their chiefess mother and also through Kamehameha V. Aunty Kai and her husband, David Ka'ae, lived next to Wai o Kamas' remaining sluice gate on Mākila beach when Mrs. Ashdown met Aunty Kai in the early 1930s. According to anthropologist Christiaan Klieger, who researched old land records for his 1998 book *Moku'ula*, as heirs of Hoapili through his bequeath of lands to Kamehameha V, the Shaw sisters might have been considered the final *kahu* (guardians) of Kiha-wahine.[99]

Ulu-mā-hei-hei Hoapili was the son of Kame'eiamoku, Kamehameha's childhood *kahu* (guardian) and trusted advisor. They grew up together and Kamehameha gave him the name Hoapili, "close companion." Kame'eiamoku and his twin brother Kamanawa were the sons of Kekaulike and half-brothers of Kahekili-nui 'ahumanu II. All Maui *ali'i* of the Pi'ilani line, Kiha-wahine had been their family *'aumakua* for ten generations. Ka'ahumanu's mother, Namahana, was a daughter of Kekaulike, and named her daughter, The Feather Cloak, after her brother Kahekili-nui'ahumanu.[100]

Eventually, the county filled in the swampland of Wai o Kama. After the death of the Shaw sisters the land was sold and in the 1970s a shopping complex and oceanfront hotel were built. Today, fresh water springs can be seen draining onto Mākila beach, between 505 Front Street and the Lahaina Shores Hotel, at the site of the old sluice gate of Wai o Kama and Mokuhinia.[101]

Since 1990, the Friends of Moku'ula, a non-profit organization, has been working towards the restoration of Mokuhinia Pond and Moku'ula. Archeological work has revealed that Moku'ula lies undisturbed beneath three feet of fill, an astonishing artifact of antiquity still considered sacred ground by many Hawaiians.[102]

Wai o Kama Fishpond. Photo by Ray Baker. R. J. Baker Collection, Bishop Museum

HAU'OLA STONE

Along the waterfront of Keawaiki (The Small Harbor) and below the grounds of the Lahaina Library, sits the Hau'ola Stone. Shaped like a seat, it was once considered sacred for its curative powers.

THE LEGEND OF HAU'OLA

Kapo Hau'ola was a *Kahuna Lapa'au* (Priestess of Medicine) in the temple of Lono, where both men and women were holders of the secrets and were adept at *Kā Hea*, The Calling Prayers, by which they performed wondrous cures. When wars began, Kapo Hau'ola devoted her life to curing her people of the ills of warfare. This angered the invading chiefs who feared her powers. Warned by her *'aumakua* that warriors were coming to kill her, Kapo Hau'ola and her assistants fled to the sea at Keawaiki. When they dove into the waters a great storm rose up and when it subsided, there stood Pōhaku Hau'ola and her many assistants, as large stones along the shore where none had been before.

Priests and people, who went to the area to remember the good Chiefess Hau'ola, found that the waters and stone had curative powers. Sick and injured people were placed on Pōhaku Hau'ola and when the final *Kā Hea* were performed, they were cured. The most notable cures were for the heart and the mind. There are several versions of the story of Hau'ola. This was Aunty Kai's version.[103]

Kā Hea is a healing ceremony that consists of five days of prayer followed by *Kapu kai*, ceremonial bathing in the sea to cleanse the mind and body. *Kapu kai* was performed before warfare and before graduation from the *Hālau Hula*, after menses for women, or contact with the sick or dead. Often *limu kala* (seaweed) was used to wash away negative energy.[104]

King Kamehameha came to Laha'āina in 1802 with his famous fleet of 800 war canoes, built in preparation to conquer Kaua'i and Ni'ihau. Known as *wa'a peleleu*, the seventy-foot long war canoes were said to stretch along the beaches from Ma'alaea to Launiupoko.[105]

Hawaiian Canoes, 1779.
John Webber. Courtesy of David Rumsey Map Collection, www.davidrumsey.com

THE BRICK PALACE

Excavated foundation site of Kamehameha I's Brick Palace, first modern building in the islands

Viewed by the Maui *ali'i* as a conquering enemy, Kamehameha was not joyfully accepted by them or their spirit ancestors. Kamehameha knew he would not be readily welcomed on Moku'ula, home of the Kihawahine, so he built a modern Brick Palace in Lua'ehu fronting Keawaiki harbor, where everyone could see who was now in charge. Said to be the first modern building in the islands, its name was even foreign. The people of Lua'ehu had no love for the Brick Palace of Kamehameha.[106]

Kamehameha brought his favorite wife Queen Ka'ahumanu, and his sacred wife Queen Ke'ōpūolani and their first child 'Iolani Liholiho, with a retinue of a 1,000 and set up his court around the new Brick Palace, far from Moku'ula. Lands were divided up and redistributed among Kamehameha's chiefs and warriors. Kamehameha made major public improvements to Lahaina, repairing taro fields and making agricultural fields productive again after the devastation of so much war.

Here in Lua'ehu, Kamehameha declares 5 year-old Liholiho as his heir, ensuring the continued rule of his dynasty. As the new head of the religion, the young prince began his ceremonial duties by touring the island and rededicating several heiau, including Heiau Wa'ilehua, to Kū-ka-'ili-moku, the war god of Hawai'i island.[107]

As Kamehameha's childhood *kahu* and trusted advisor, Kame'eiamoku, lay dying in Lahaina, he told Kamehameha that Kahekili had been his real father and gave him the tokens of proof. Kamehameha wept for the loss of his brothers, but Kame'eiamoku told him that only by living his destiny could he win a lasting peace and end centuries of warfare.

It seems the beautiful Princess Keku'iapoiwa II, niece of King Alapa'inui of Hawai'i, was visiting the court of King Kahekili in Wailuku when the two fell in love. Princess Keku'iapoiwa II became *hāpai* (pregnant) but had been betrothed to Prince Keoua-nui of Hawai'i since childhood. Her uncle demanded she come home and marry her betrothed, and that the infant be destroyed when born. To avoid all out war, the princess returned home and when the infant was born, her priest Nae'ole stole him away with her consent, hiding him in the high country of 'Āwini Falls, deep in the head of Po-lo-lū Valley for five years. King Kahekili-nui-'ahumanu had named the boy Kamehameha after his brother Kamehameha-nui, who had bequeathed the Maui realm to him.

When the son of an *ali'i* reached five years of age, the custom was to present him at court. By now, Alapa'inui was old and had forgiven his niece, who had married Keoua-nui. Alapa'inui accepted young Kamehameha, who was generally known as the son of Keoua-nui, but Alapa'inui also accepted his name, Kamehameha, after Kahekili's brother Kamehameha-nui, and also the twin half-brothers of Kahekili, Kame'eiamoku and Ka-ma-na-wa, sent by Kahekili as *kahu* (guardians) for the young prince.[108]

Leaving his close friend Hoapili as Governor of Maui after two years at Lahaina, Kamehameha moved his court and fleet of war canoes to O'ahu, still anticipating an invasion of Kaua'i. Here he made altars to Kiha-wahine and worshipped her among his many gods.

In 1810, King Ka-umu-ali'i ceded Kaua'i and Ni'ihau to Kamehameha's rule, finally uniting the entire group of islands as the Kingdom of Hawai'i.[109] In 1812, Kamehameha returned his court to his home island of Hawai'i where Ke'ōpūolani gave birth to the sacred *ali'i* Kau-i-ke-ao-uli (Kamehameha III) in 1814 and the sacred Princess Nahi'ena'ena in 1815.[110]

By 1795, an estimated 50 ships of English and American origin had made port in Hawai'i for provisions. Kamehameha took advantage of western trade and commerce and had many western advisors, but he held strict to the old religion and resisted all offerings of Christianity. No trade was allowed, by anyone, without Kamehameha's permission.

Lahaina from the anchorage. Benson Lossing and William Barrett, engravers, From *Sandwich Island Notes by* George Washington Bates, Harper & Brother, 1854

In 1811 Kamehameha signed an agreement with Boston ship captains and established a monopoly on sandalwood trade to China, where the fragrant wood was made into incense. Kamehameha controlled the trade by not allowing any credit and by placing a *kapu* on young trees.

As more trade ships stopped in Hawaiʻi, sandalwood was used for the acquisition of Western goods. Kamehameha bought the brig *Columbia* for two shiploads of sandalwood.

After Kamehameha's death, Kamehameha II became indebted to sandalwood traders. Soon chiefs began pillaging the sandalwood forests, and by 1830 they were exhausted of the precious wood and its economic benefits.[111]

KING KAMEHAMEHA I

Kamehameha, 1816.
Louis Choris. From *Voyage pittoresque autour de monde,* F. Didot, Paris, 1822

Kamehameha the Great died peacefully at his old home in Kailua-Kona at the age of 87 in May of 1819. Princess Nahiʻenaʻena was just four years old and her brother Prince Kauikeaouli was six. Kamehameha's bones were entrusted to his closest companion, Hoapili.[11]

By the time of Kamehameha's death, in just 40 years since western contact, it was estimated that 50 percent of the Native Hawaiian population had perished due to western diseases.[113]

KING KAMEHAMEHA II

Kamehameha II, 1824.
John Hayter, National Library of Australia, nla.pic-an9897032

At the Council of Chiefs, twenty-two year-old ʻIolani Liholiho was proclaimed King Kamehameha II and Queen Kaʻahumanu was granted co-rulership as *kuhina nui*, at her insistence, claiming that Kamehameha had willed this to her.[114] No one questioned her authority and Liholiho agreed to this new form of "shared" rule with his imperious step-mother Queen Kaʻahumanu. As requested by Kamehameha, at his death, Hoapili became guardian of sacred Keʻōpūolani and her children. This also made him guardian of her *ʻaumakua*, Kihawahine.[115]

ʻAI KAPU

Queen Kaʻahumanu, 1816. Louis Choris. Hawaiʻi State Archives

Of all the *kapu* (taboos), the most discriminatory to commoner and *aliʻi* alike was the *ʻai kapu* (*ʻai* – food, *kapu* – forbidden). This prohibited men and women from eating together and made certain foods forbidden to women, such as pork and bananas. Contact with Westerners over the past forty years had already undermined faith in the *kapu* system, as Hawaiians observed these foreigners breaking sacred *kapu* without the promised punishment. Nonetheless, to disregard this restriction would unravel all *kapu* and the ancient religion with it.

As was the custom during the period of mourning, all restrictions were lifted at Kamehameha's death, including the *ʻai kapu.* Instead of reinstating it, as was traditionally done by the new king, Queen Regent Kaʻahumanu, her sister Ka-hei-hei-malie, and Liholiho's sacred mother, Dowager Queen Keʻōpūolani, eventually persuaded the young king to eat with them in public. These powerful matriarchs, all wives of his father, wanted the cruel sacrificial punishments for broken *kapu* to stop. *Nīʻau piʻo* Queen Keʻōpūolani had the highest rank and therefore the strictest restrictions which kept her isolated much of the time. By breaking this centuries old religious *kapu*, these *aliʻi* women knew this act of defiance would undermined the entire *kapu* system and its *Kānāwai Akua* (Laws of God).[116]

In 1820 Liholiho, now King Kamehameha II, accompanied his seven year-old

Princess Nahi'ena'ena's ali'i-*style* hale pili, *Halekamani, in Lua'ehu.* Bishop Museum

brother Kauikeaouli and five year-old sister Nahiʻenaʻena and their attendants to Lahaina, where the Prince and Princess were established in separate residences in Luaʻehu.[117]

Into this confusion the tall ships of the New England whalers had begun arriving in Lahaina for provisions in 1819. Between 1820 through the 1860's, Lahaina Roadstead, as the ʻAuʻau channel was renamed by sailors, became the principal anchorage of America's whaling fleet. As many as 1,000 whalers could be on the streets of Lahaina at one time, overwhelming the little town of 3,000. In 1846, at the peak of whaling, Lahaina was visited by 596 whaling ships in one year. [118]

Soon Lahaina became an economic center for provisioning whalers and the

Old Lahaina Roadstead, 1843. Pioneer Inn

export of Hawaiian sandalwood to China. The sailors used Lahaina for rest and recreation, debasing the women and cavorting drunk in the streets.

Much to their disappointment, Calvinist missionaries arrived right behind them from their own New England home, bringing with them the puritanical values the sailors thought they had left behind Cape Horn.

In March of 1820 the first party of missionaries arrived off Kona in the ship *Thaddeus*. Among the party of missionaries were four native Hawaiians. They had found their way to New England as young sailors and managed to receive an education at the Foreign Mission School in Connecticut. They were invaluable in convincing King Kamehameha II and Queen Regent Ka'ahumanu to allow the missionaries to stay and preach their gospel.

HENRY 'ŌPŪKAHA'IA

First Hawaiian Convert to Christianity

Henry 'Ōpūkaha'ia, c. 1817. Hawai'i State Archives

The inspiration to bring Christianity to the Sandwich Islands came from a native son. Brutally orphaned in the wars of Kamehameha, Henry 'Ōpūkaha'ia left his home on Hawai'i island as a young teenager aboard a merchant ship around 1807. Eager to learn, he was brought to Connecticut by the ship's captain in 1809. 'Ōpūkaha'ia was educated by various members of the American Board of Commissioners for Foreign Missions, and was a student in the first class of the Foreign Mission School in 1817 with the four other Native Hawaiian students. 'Ōpūkaha'ia worked with missionary Samuel Ruggles to develop grammar and an alphabet for the Hawaiian language and then he translated the *Book of Genesis* into Hawaiian. ʻŌpūkahaʻia wrote the memoirs of his early "pagan" life and implored his Protestant teachers to bring the gospel to the people of his "heathen" homeland.

ʻŌpūkahaʻia died suddenly from typhoid fever in 1818 at the age of 26 years. His mentor, E. W. Dwight, compiled and published *Memoirs of Henry Obookiah* that same year and it was used to raise funds for the mission and to inspire Protestant missionaries to take up the cause of the "heathens" in the Sandwich Islands. Within a year the *Thaddeus* set sail with a physician, a teacher, a farmer, two ministers and their brave wives, and the four native Hawaiians. They brought with them ʻŌpūkahaʻia's Hawaiian *Book of Genesis* and used it to teach the missionaries the Hawaiian language during their long voyage.[119]

When they arrived in Kona, Hawaiʻi, they were surprised and elated to find that the established religion had been overthrown and that temples and images were being destroyed. It was perfect timing for the acceptance of a new religion.

Queen Kaʻahumanu resisted the puritanical ideas of the missionaries, who seemed to want to replace the old *kapu* she had just abolished, with new *kapu* that forbade cultural traditions like *hula* and plural marriage.[120]

After the death of Kamehameha, Kaʻahumanu married King Kaumualiʻi of Kauaʻi, and for good measure also married his son, thereby securing the allegiance of Kauaʻi's chiefs.[121] In May and June of 1822, Kaʻahumanu and Kaumualiʻi toured the islands, feasting, dancing, burning idols, and destroying temples.[122]

Thomas Hopu, Prince George Kaumualiʻi, Willliam Kanui, John Honoliʻi.
From a painting by Samuel Morse, 1816. Mission House Museum

QUEEN KEʻŌPŪOLANI

No known photograph or drawing exists of Keʻōpūolani. This painting by David Parker was commissioned by Kamehameha Schools in 2001.
Used with permission from Kamehameha Schools

Queen Mother Keʻōpūolani, however, was inspired by the story of Genesis, read to her by the Reverend Richards, because it so closely resembled the *Kumulipo*, allegorically, a manner of speech very familiar to an *aliʻi* of her rank. The *Kumulipo* was the basic pillar of God and Aloha to Hawaiians, centuries prior to the arrival of Europeans. Keʻōpūolani now wanted to learn to read and write her own language.[123]

At the request of an ailing Keʻōpūolani, the Reverends Charles Stewart and William Richards and their families accompanied the Queen and her husband Hoapili, the newly appointed Governor of Maui, to Lahaina in May of 1823, where she assisted them in every way to establish their Mission. Eager to learn to read, Keʻōpūolani is Reverend Richards' first pupil and quickly learns to read the *palapala,* (the word of God). All her chiefs and the court are instructed in reading, writing and religion.[124] Her husband, Governor Hoapili, and Queen Keʻōpūolani become early proponents of this new religion.

On September 16, 1823, noble Keʻōpūolani is baptized and takes the Christian name Harriet after her friend and teacher, the wife of Reverend Stewart, shortly before passing away at the age of 47 years. She died surrounded by her royal family and the Reverends Stewart and Bingham. She had asked her teacher, Reverend Richards, to educate her children and to counsel Princess Nahiʻenaʻena in the Christian life.

Funeral procession of Queen Ke'ōpūolani, 1823.
William Ellis. Hawaiian Mission Children's Society

At her instructions, no ancient customs (such as self-mutilation), except wailing, were allowed after her death. Most sacred Ke'ōpūolani, last of the *nī'au pi'o ali'i*, received the first Christian burial of an *ali'i* in Hawai'i, at Moku'ula, home of her *'aumakua*, Kiha-wahine.[125]

Within days the whole district of people, young and old, *ali'i* as well as commoner, came together for a week to pass the stones of Heiau Wa'ilehua at Mākila Point along the shore for more than a mile to encase their Queen's tomb in coral stone. [126]

Coral tomb, Lahainaluna engraving.
Lorrin Andrews. From *1840 Book of Sermon*

KING KAMEHAMEHA II AND QUEEN KALAMA

King Kamehameha II and Queen Kamāmalu and suite at Drury Lane Theatre, 1824

Two months after Queen Dowager Ke'ōpūolani's death, Liholiho, King Kamehameha II, and his Queen Kamāmalu depart with their suite for London to meet with the King of England to learn about the British form of monarchy. Liholiho named his brother Prince Kau-i-ke-aouli as his heir and Queen Ka'ahumanu as regent before departing in November 1823. Sadly, both Liholiho and Kamāmalu succumbed to measles and died in London in July 1824, before they could meet King George IV.[127]

KING KAMEHAMEHA III

King Kamehameha III, 1825. Robert Dampier. Honolulu Museum of Art

Missionary preaching to Hawaiians, 1838. National Library of Australia

Young Prince Kauikeaouli, at 12 years of age, becomes King Kamehameha III and at Kaʻahumanu's direction, he moves his court to be with her in Honolulu.

Princess Nahiʻenaʻena stays with her stepfather, Governor Hoapili, at Luaʻehu and continues her Christian studies with Reverend Richards.[128] After Keʻōpūolani's death, Governor Ulumāheihei Hoapili marries Kaʻahumanu's sister (and also a widow of Kamehameha I) Kaheiheimalia, in the first Christian marriage of *aliʻi*. She thereafter took the Christian name, Hoapili Wahine (Mrs. Hoapili).

Struck by the sudden deaths of her two adopted children, Liholiho and Kamāmalu, and her beloved husband Kaumualiʻi earlier that year, Queen Regent Kaʻahumanu asks to be instructed in the Christian life. Like Keʻōpūolani, she is intrigued by the art of reading and writing and soon surrenders fully to the Christian faith. By September of 1824 she is commanding all the chiefs and *aliʻi* to learn the *palapala* (writing) and *pule* (prayer). By the end of the year, nine year-old Princess Nahiʻenaʻena and her stepfather Hoapili had instructed 270 children.[129]

The bodies of King Kamehameha II and Queen Kamāmalu were returned to Lahaina in May 1825 on the *HMS Blonde* of the British Navy. Both received Christian burials on Oahu.[130]

Site of first Mission home of the Reverend William Richards and wife Harriet

Queen Ke'ōpūolani had given Charles Stewart and William Richards their choice of land near the beach in Lua'ehu for homesites. Stewart soon returned to the United States with an ill wife, but Richards and his wife lived in a *pili* (grass) house at this site on Front Street for four years before building the first coral stone house in the islands made with mortar and plaster. Consisting of two stories with verandas, it also had a cellar. Shortly after construction, it was in this cellar that the missionary families took refuge during the cannonading by the English whaler *John Palmer* on October 27, 1827, protesting the new restriction on women visiting the ships. [131] Later, in 1834, the missionary Ephriam Spaulding built a similar house next door that came to be known as the Baldwin House.

LUA'EHU LAWS

On December 5, 1827, the first written laws of the Kingdom were proclaimed against murder, theft, fighting, and desecration of the Sabbath. Written by the Reverend Richards, Queen Ka'ahumanu, Governor Hoapili, and young King Kamehameha III, they were founded on the Ten Commandments and the *Kānāwai Akua* (Laws of God). Written in the thatched long house near the Brick Palace, these first laws were known as the Lua'ehu Laws for where they were written.[132]

WAINE'E CHURCH

Waine'e Church, 1909.
Photoprint by J.J. Williams. DOE Oceania: Amer Polynesia: Hawaiian: NM 50997 04896500, National Anthropological Archives, Smithsonian Institution

The first church built on Maui was a thatched long house with six windows and a doorway, located on what is now Waine'e Street. Built on land given by Governor Hoapili, who lived on the adjacent land, the church was dedicated as "Ebenezer" at Hoapili's request in April, 1828.

In 1832 a new church was dedicated on the site. Laid out by Reverend Richards, it was the first stone church built in the islands. It resembled what Kawaiaha'o Church in Honolulu would become and was called Wai-ne'e (moving water) after the name of the district through which the moving waters of Kaua'ula Valley flowed, feeding the fields and the fishponds on its way to the sea at Mākila beach. The large church seated three thousand. Queen Ka'ahumanu gifted the bell for the church tower, which is still on display today in the churchyard.[133]

Ali'i were paddled across Mokuhinia Pond in canoes, resplendent in their feather capes, to attend services at Waine'e Church.

In 1858 the fierce Kaua'ula Wind took off the steeple and half of the roof. It took considerable time to make repairs as it was necessary to send to the Puget Sound for timbers.[134] In 1888, Aunty Kai's *ali'i* mother, Lahela Ma'ele Napuhiwa-Ha-pa'u Shaw, formally deeded the land of the Waine'e Cemetary to the Church.[135]

Royal tombs

The tombs of Queen Ke'ōpūolani, her daughter Princess Nahi-ena'ena, and her husband and *kahu* Governor 'Uluma-hei-hei Hoapili, as well as King Kaumuali'i of Kaua'i (Ka'ahumanu's husband) and their Christian teacher, the Reverend William Richards, rest at Waine'e Cemetery. Some believe the remains were moved to the Royal Mausoleum in Nu'uanu, O'ahu.

On June 24, 1894, anti-Royalists burned the church down. H. P. Baldwin, a son of missionary Dwight Baldwin and a staunch royalist, rebuilt the church in 1897. In 1951, old Waine'e church was destroyed completely by the Kaua'ula Wind. Rebuilt of cement blocks in 1953, Waine'e (Moving Waters) Church was rededicated as Waiola (Waters of Life) Church.[136]

WAIOLA CHURCH

HALE ALOHA

From *The Centennial Book: One Hundred Years of Christian Civilization in Hawaii, 1820–1920* by the Central Committee of the Hawaiian Mission Centennial, ed., 1920

In 1823 a *pili hale hālawai* (grass meeting house) was built at this site and used for church meetings, prayer meetings, and school classes. In 1858 Hale Aloha was built here under the direction of Dr. Baldwin and the sponsorship of Governor Hoapili. Constructed between 1855 and 1858, this coral-stone church was also used as a schoolhouse for many years. A noted principal, Henry Dickenson, had a street in Lahaina named after him.[137] The Hawaiians named the church Hale Aloha (House of Love) in thanksgiving of Lahaina's escape from the small pox epidemic of 1853.[138]

Restored in 1974 by the County of Maui, Hale Aloha still stands to the rear of the Episcopal Cemetery on Waine'e Street, under the management of Lahaina Restoration Foundation.[139]

Hale Aloha today

BALDWIN HOUSE

This Mission house was built in 1834 by Reverend Ephraim Spaulding. In 1835, Dr. Dwight Baldwin and his family were transferred by the Mission to Lahaina from Waimea, Hawai'i, and stationed in this house. Dr. Baldwin added two rooms in 1840 to include a dispensary, and in 1847 and 1848 he added a second story to accommodate his growing family of six children.[140]

Dr. Baldwin was credited with saving Maui, Moloka'i, and Lāna'i from the small pox epidemic of 1853 that devastated parts of O'ahu and Hawai'i and killed over 3,000 people. Hale Aloha church was built in commemoration and thanksgiving of that event.

Dr. Baldwin's son, Henry Perrine Baldwin, grew up to become a pioneer of Maui's sugar industry. He would eventually partner with another missionary son, William Alexander, his brother-in-law, and they would build the Alexander and Baldwin sugar dynasty that until today is the largest single private landowner on Maui.[141]

This original building was restored in the 1940s by H. P. Baldwin's son Frank at the request of Mrs. Ashdown and the West Maui Hawaiian Civic Club.[142]

LAHAINALUNA SEMINARY

Lahainaluna, upper left, 1831.
From *A Residence of Twenty-one Years in the Sandwich Islands*
by Hiram Bingham, H. Huntington, 1847

Governor Hoapili and his wife Hoapili Wahine gave 1,000 acres of the uplands of Lahaina to the Mission to build a school for the training of adult male students to assist in the field of teaching. Beginning in September 1831 with two grass *hale*, Lahainaluna (Upper Lahaina) Seminary became the first school west of the Rocky Mountains. That first year produced sixty students, all adult males.[143]

Under the direction of their principal Reverend Lorrin Andrews, the students completed the first stone schoolhouse in June 1832. That same year the school received an old Ramage printing press from the Mission in Honolulu and immediately began printing Hawaiian language texts in a grass *hale* (house).

By 1841, public education had grown to the point of needing an administrator. One of Lahainaluna's first graduates, David Malo, became Superintendent of Schools for Maui, Lāna'i, and Moloka'i. He reported 10 schools, 10 teachers and 537 scholars in Lahaina in April 1841.[144]

By 1850, Lahainaluna had sent out over 400 educated Hawaiians who became leaders in teaching and government service. Subjects from algebra to theology were taught. By now, Hawai'i boasted the highest literacy rate in the world. At the 1869 Pan-Pacific Exhibition in Paris, Hawai'i exhibited over twenty-two textbooks, many newspapers and manuscripts and the Bible, all in the Hawaiian language. Europeans were shocked to see the advanced learning of the so-called "pagan" Hawaiians, who had enjoyed a constitutional monarchy since 1840. Europeans were still ruled by absolute monarchies and schooling for the common people of America and Europe was not yet available.[145]

Among the most well-known early graduates were David Malo and S. M. Kamakau, both amateur historians who wrote important books on ancient Hawaiian culture that are still referenced by researchers today.

HALE PAʻI

Photo by Joel Bradshaw

Built in one year by the students themselves, Hale Paʻi (Print House) was made of coral and lava stones plastered in cement. By February 14, 1834, the first newspaper ever printed west of the Rockies was written and printed in Hawaiian by students at Lahainaluna. It was called *Ka Lama Hawaiʻi* (Torch of Hawaiʻi), said by Principal Andrews to be chosen because "knowledge is like light."

Textbooks were illustrated with engravings made on copper sheets obtained from ships captains, who carried them to reinforce their hulls against damage from tropical organisms.[146] Many of the sketches for the etchings were drawn by the artist missionary Edward Bailey of the Wailuku Female Seminary.[147] His paintings of Maui can still be viewed today at the Bailey House Museum in Wailuku.

Hale Paʻi printed the first Hawaiian language Bible, the first written laws of the Kingdom, the first Hawaiian Constitution, the first Hawaiian language newspaper, and Hawaiʻi's first paper money. By 1840 there were 22 textbooks in the Hawaiian language.

Hale Pa'i, c. 1915. Photo by Ray Baker.
R. J. Baker Collection, Bishop Museum

WAI O WAO

Hale Paʻi (Print House) was built next to Wai o Wao, the ancient *'auwai* (irrigation ditch), that ran from the headwaters of neighboring Kahana Valley. Named for the Kiha ancestress, Wao, who lived in Kahana Stream, Wai o Wao also became her home. Mrs. Ashdown said the old folks told stories of young men being lured into the valley by the voice of Wao.[148]

By 1847, to meet the needs of growing agricultural interests, the 'Auwai-o-Piʻilani from Kauaʻula Valley and the 'Auwai Wai o Wao were modified into the Lahainaluna Ditch.[149] This allowed the students to plant much of the surrounding land in taro, bananas and sugar cane, which they sold to the local mills.[150]

Hale Paʻi is the only original building of Lahainaluna School to have survived. Its distinctive New England-style high-pitched gable roof with wooden shingles covers a two-story building with a full basement.[151] In 1982 Hale Paʻi underwent a two-year restoration effort, with funding from the State of Hawaiʻi. It serves today as a museum managed by the Lahaina Restoration Foundation.

PĀPELEKĀNE

Replica of a section of the Old Fort

In order to quell the outlandish behavior of drunken sailors, Queen Ka'ahumanu came to Lahaina in 1831 to assist Governor Hoapili in the construction of a fort fronting the harbor. All of her subjects who had come to pay homage to their Queen helped to build the 12 foot thick walls made of coral stone cut from the reef. It was completed in one month.[152]

A guard in a *malo* (loincloth) paced the wall at sunset and blew a *pū kani* (shell trumpet) to warn sailors to return to their ships. Any crewmen caught after sunset were locked in the fort until daylight. The fort came to be known as Pā-pelekāne. (*Pā* – tied, *pelekāne* – Britisher)[153]

A roster of fort personnel made in 1848 listed 46 soldiers, 9 officers, and 5 drummers.[154] The fort was demolished and used for building materials on the prison in 1852. The portion of the fort on display today is a replica of the original walls.

QUEEN REGENT KA'AHUMANU

Ka'ahumanu, 1816.
Louis Choris. Honolulu Museum of Art

Queen Regent Ka'ahumanu died the followng year in 1832, at the age of sixty-four. She was at her country home in Mānoa Valley, O'ahu and was buried in Nu'uanu after a service at the thatched Kawaiaha'o Church. In eight short years the powerful Ka'ahumanu and her Calvinist missionaries had replaced the *kapu* of the ancient religion with *kapu* of a new religion, *kapu* on alcohol, *kapu* on sibling marriages, *kapu* on plural marriages, and *kapu* on sex without marriage.

Wise to the ancient ways of the *ali'i nui*, Kauikeaouli Kamehameha III used this opportunity to reinstate his inherited rights by breaking the *kapu* of Ka'ahumanu. For a brief time, liquor, tobacco. and the banned hula came back. Known as the Kings Rebellion, Kauikeaouli was just 19 years old when he sent word from O'ahu that he would appoint the anti-Christian royals Liliha as *kuhini nui* (Premier) and Governor of O'ahu and Abner Pākī to assist her. Governor Hoapili immediately rushed to O'ahu from Lahaina and confronted the young King. Kauikeaouli gave in to his stepfather and the rebellion was crushed. The King's half-sister, the very pious Christian Kīna'u, was appointed Kuhina-nui to replace Ka'ahumanu.[155]

Premier Kīna'u, 1836.
Barthélémy Lauvergne. Honolulu Museum of Art

PRINCESS NAHIʻENAʻENA

Princess Nahiʻenaʻena, 1825.
Robert Dampier. Honolulu Museum of Art

THE KING AND THE PRINCESS

Kamehameha III and his sister Princess Nahiʻenaʻena were devoted to each other, having been raised together as sacred *aliʻi* and having suffered great losses together. Kauikeaouli wanted a *piʻo* marriage with Nahiʻenaʻena to perpetuate their divine lineage. He saw this as his traditional right, but the missionaries and his Christian relatives were shocked and forbade the union.

Nahiʻenaʻena tried very hard to live the Christian life as her mother had asked, devoting much of her time to religious duties and touring the island with Reverend Richards to encourage the *palapala* (the word of God). The people worshipped her wherever she went.[156]

But the love for her brother and the festivities of his traditional court in

Honolulu kept Nahi'ena'ena going back and forth between two different worlds, one as an *ali'i nui* of her brother's court, and the other as a missionary of the new gospel. In the end, she chose her brother and in 1834 they were united in a *pi'o* (brother-sister marriage) ceremony before the chiefs of 'Ewa.[157]

Perhaps in an act of repentance, Nahi'ena'ena returned to Lahaina in 1835 and was "married off" to her nephew Leleiohoku, a grandson of Kamehameha I, in a Christian ceremony performed by her mentor Reverend Richards at Waine'e Church.

Still, she returned to Honolulu and became pregnant in early 1836. Later that year she gave birth to a premature child who died shortly thereafter. Beautiful Princess Nahi'ena'ena never recovered. She died on December 30, 1836 at just 21 years of age.[158]

The King was grief stricken and he retreated from Honolulu to the sanctuary of Moku'ula, still *kapu* amid his rapidly westernized world. In early 1837 he brought Nahi'ena'ena's body to Lahaina and placed her temporarily in her mother's tomb until a stone house could be completed on Moku'ula. Kauikeaouli moved his mother and beloved sister and their child to the stone tomb on Moku'ula and then took up permanent residence on the sacred island. A sanctuary from all that the western world had brought to his homeland, Moku'ula was still a symbol of the old religion.[159]

Shocked and sobered by Nahi'ena'ena's death, Kamehameha III pledges himself to sobriety. In February 1837, Kauikeaouli married Kalama Kapakuhaili, a chiefess of Hawai'i, and they take up residence on Moku'ula, where King Kamehameha III and Queen Kalama will live for the next eight years.[160]

Princess Nahi'ena'ena, October 1836.
Barthelemy Lauvergne. Bishop Museum

SEAMAN'S HOSPITAL

Located at the northern end of Lahaina, far from Lua-ʻehu where the *aliʻi* and missionaries lived, a young Kamehameha III secretly conspired with a Chinese merchant to build a rooming house and saloon for ship captains on this property in 1833. Away from the judgmental eyes of his stepfather Governor Hoapili and his half-sister Premier Kīnaʻu, both devout Christians, as well as his advisor and former teacher Reverend William Richards, the 20-year-old King could escape the restrictions of the new religion and indulge his fun loving nature in cards and drink here.

After the tragic death of his beloved sister Nahi'ena'ena in December 1936, Kauikeaouli took a pledge of sobriety and closed the saloon.

Sometime in 1841, Kamehameha III gave the building over for the care of sick and injured seamen. At the King's request, the American Government sent agents and doctors to operate a United States Marine Hospital to care for disabled American seaman. Known as the Seaman's Hospital, it housed as many as 100 sailors at a time in the four small wards that were once behind the hospital. All manner of illness and mishap were treated. In a visit to the hospital in 1853, Reverend Sereno Bishop recalled in his journal, "...a fine looking young man, full attention, whose legs are shackled with heavy irons to prevent his sudden escapades. He was struck on the head by a whale last winter and is deranged. When his fits are on, the place resounds with his jolly songs."

Operating for twenty years during the peak of the whaling economy, Seaman's Hospital closed in September 1862 due to the decline in whaling ships calling at Lahaina. Thereafter, sick seamen were sent to Honolulu.

In 1864 the Episcopal Church acquired the building and the Sisters of the Holy Trinity established St. Cross School, which operated until 1877.[161]

Having fallen into disrepair, the site was purchased by the Lahaina Restoration Foundation in 1975 and the building was completely reconstructed in 1982.

Anchor from whale ship

SEAMAN'S CHAPEL & READING ROOM

In hopes of providing a place of respite to sailors far from home and to also remind them that God had not been left behind at the Horn, the Mission sought funds from ship captains to help fund the building of the Master's Reading Room. Completed in 1834 by the Reverend Ephriam Spaulding, it is made of coral stone, lava rock, and mortar. It had a basement for use by the missionaries, the lower floor as a reading room, and the second floor was used as a chapel and vestry.[162] Located next to the Baldwin House on Front Street, this building was completely reconstructed by the Lahaina Restoration Foundation in 1982.

HALE PĪʻULA

Site of Kamehameha III palace, Hale Piʻula

Once the site of the palace of his ancestor the great King Piʻilani, Kamehameha III chose this strip of land between Mokuhinia Pond and the beach to build his formal Western-style palace named Hale Pīʻula (House of Sacred Blessing). This is a contraction of *Pī-kai*, to bless, and *ʻula*, sacred. Pī-kai and Paʻa ʻula are names for blessings with salt water.[163]

Built of stone and imported wood, it was a two story building surrounded by a covered lanai. Never completely finished, the King used Hale Pīʻula only for official business.

Hale Pīʻula served Kamehameha III as the site for the signing of the first western laws to govern his Kingdom. With the assistance of Reverend Richards, who had resigned the Mission to become full time advisor to the King, the first Bill of Rights was adopted in 1839. The Edict of Toleration, signed on June 17, 1839, gave religious freedom to his people and freed Catholics from persecution by Calvinist missionaries and their *aliʻi* supporters.

Kamehameha III signed the first written Hawaiian Constitution on October 10, 1840, changing the government of the Hawaiian people from an absolute monarchy to a constitutional monarchy, consisting of the ruling monarch (executive branch), a two house legislative branch: one of the *aliʻi* and one an elected assembly, plus a judiciary. This new government of the Hawaiian Kingdom had its first Legislative Session at Hale Pīʻula in 1840.[164]

In 1842 the noble Governor Hoapili and his wife Kaheiheimalia died, as well as Kamehameha III's half-sister Premier Kīnaʻu, Kaikioewa, the Governor of

Kauai, and the great Kapiʻolani, who defied the goddess Pele at the fires of Kīlauea volcano. These *aliʻi* were among the earliest converts to Christianity and the death of these five nobles was a great loss to the monarchy and the Mission, as none left any children to fulfill their leadership positions, except Premier Kīnaʻu, whose sons Alexander Liholiho and Lot Kamehameha would become Kamehameha IV and V, respectively.[165]

By 1845 Honolulu had become the center of commerce with its deep-water port and Kamehameha III was forced to move his court to Oʻahu. Many *aliʻi* and their descendants continued to maintain residences in Luaʻehu for years to come.[166]

In 1848 the *Great Māhele* (to divide or portion) was enacted to provide for title to private property. Kamehameha III was convinced by the sugar planters that this was necessary for Hawaiʻi to participate in a western economy. Previously, private land ownership did not exist. Commoners worked the land collectively and paid taxes to the chief of their *ahupuaʻa* (land division), who ruled at the privilege of the King.

The *Great Māhele* divided the land into thirds; one-third to the Crown as Crown lands, one-third to the chiefs, and one-third to the common people. However, this required the commoner to file written claims for the land they worked within two years. This was not clearly understood by the populace and at the end of the two-year period, only one percent of the land had been claimed. Eventually it was sold to foreigners.[167]

In 1848 Hale Pīʻula became a courthouse and was in use until it was destroyed by the Kauaʻula Wind in 1858. In 1859 the rubble from the palace was used to construct the Lahaina Courthouse on the previous site of the old fort fronting the harbor.[168] Today the commemorative site is a county park named Kamehameha Iki. Now used by canoe clubs, traditional cultural practices and sacred blessings still take place on this royal site.

Lahaina, West Maui, Sandwich Islands (part), 1855.
James Gay Sawkins, nla.pic-an3018524-3-v. National Library of Australia

HALE PA'AHAO

Stockade wall of Hale Pa'ahao, Waine'e Street

In 1850 King Kamehameha III codified and formalized the laws of the Kingdom into a new western legal system called the Penal Code of the Hawaiian islands. Also called the Puritanical Blue Laws, these were the first printed code of laws listing crimes and punishment. Anyone causing disturbance after dark could be fined $10 and if the fine were not paid, it meant a year in jail. Reckless riding about was fined from $5 to $500. Adultery was $30 or eight months in prison.

In 1851, an "Act Relating to Prisons, their Government and Discipline" was passed by the Legislature and approved by the King, authorizing a new prison for Lahaina. By 1854 Hale Pa'ahao (Stuck-in-irons) had been completed, with wooden cell buildings and a thick coral block wall surrounding a stockade. Built from the walls of the old fort, its entry supported a gatehouse where the prison master and his family lived. In 1855 there were 330 arrests for drunkenness, 88 for assault and battery, and 111 arrests for adultery and fornication.[169]

In 1967, the Lahaina Restoration Foundation restored Hale Pa'ahao and maintains it as a museum.

KEAWAIKI LIGHTHOUSE

As the number of visiting ships increased, the provisioning of ships became big business. Trading was tightly controlled by the chiefs, who licensed the sellers. At one time, a market place was provided along the stream canal that went from Keawaiki to the spring behind Reverend Richard's house, where fresh water was obtained for visiting ships.[170]

In 1833 at 18 years of age, Princess Nahiʻenaʻena issued an edict regulating trade at the market place where women were forbidden from going "for the purpose of sightseeing or to stand idly by." It said in part, "These are the things which I strictly forbid; overcharging, under-selling, illicit selling, wrangling, breaking of bargains, enticing, pursuing, chasing a boat, greediness, and every other thing that will make a sale faulty."[171]

As the number of ships increased, so did the need for safe anchorage. In November 1840, Kamehameha III ordered the construction of the first lighthouse built in the Hawaiian Islands. Approximately nine feet tall and built of wood, the structure had two lamps for guiding ships entering the narrow channel in the reef. In 1866 it was raised to 26 feet.[172] Today's lighthouse was built by the U.S. Coast Guard in 1916.

MARIA LANIKILA CHURCH

The first Roman Catholic priests arrived from France in 1827 and were not welcomed by the Protestant missionaries or their Christian *ali'i* and were eventually banned by Ka'ahumanu. After Kamehameha III signed the Edict of Toleration in 1939, giving his people religious freedom, the Roman Catholics came to Lahaina in 1841 and built a chapel up in Kaua'ula Valley and at the present Maria Lanikila church site.[173]

The Church of the Sacred Hearts of Mary and Jesus was established in 1846 by Father Aubert Bouilloon. Maria Lanakila, Our Lady of Victory, was officially dedicated September 8, 1858. Starting with an adobe building and grass hut chapels, the stone church that stands today was completed in 1873. In 1918 it was enlarged while retaining the original structure within the interior of the church.

Maria Lanikila interior

After almost 30 years of rule, longer than any monarch of the Hawaiian Kingdom, Kauikeaouli Kamehameha III died on December 12, 1854. He had successfully ushered his country into the modern 19th century. His nephew and heir, Alexander Liholiho, succeeded him as Kamehameha IV.

Kamehameha III.
Painted by unknown artist. Bishop Museum

HOLY INNOCENTS EPISCOPAL CHURCH

Brought to Hawai'i in 1862 by King Kamehameha IV and Queen Emma, the Church of England, or the Episcopal Church, was established on royal lands in Lua'ehu in 1863. Named Holy Innocents Church, it first stood where Kamehemeha III School now stands, once a royal compound of *ali'i* residences surrounding Hale Pī'ula. In 1874 a new church was built at the corner of Front and Prison streets, now a county parking lot. The current site, next to its original, was acquired in 1908 and the current Holy Innocents Church was built in 1927.

The stained glass windows above the altar are from the first and second church.

KING KAMEHAMEHA IV AND QUEEN EMMA

Composite of Kamehameha IV and Queen Emma, c. 1860

Holy Innocents interior

The paintings on the altar depict the Blessed Mother and Child as Hawaiians and the Hawaiian Trinity—Kāne, Kū, and Lono. The Hawaiian birds are indicative of the Holy Ghost. The murals were painted prior to the war in 1941 by DeLos Blackmar, a New York artist. *Kāhili* (Royal Standards) flank the altar. [174]

Long ago the Heiau Hale Kumu-ka-lani once stood on this site. *Kumu* is foundation or pillar, *ka lani*, in this case, refers to the Heavenly Ones, the *ali'i* who lived here. [175]

Kamehameha IV and Queen Emma kept a home here. Chief Abner Pākī and his wife Konia had a home here with their daughter Princess Bernice Pauahi Pākī and their adopted daughter, Princess Lydia Ka-maka-'eha Kapa'akea, who would become final heir to the throne, Queen Lili'uokalani.[176] All had homes *makai* (shoreward) of the present Holy Innocents Church.[177]

Bernice Pauahi Pākī and Lydia Kamaka'eha Pākī (Lili'uokalani) c. 1859. E. F. Howard, used with permission from Kamehameha Schools

PIONEER MILL COMPANY

Pioneer Mill smokestack

In 1850, Irish immigrant James Campbell arrived in Lahaina where he set up a carpentry business. In 1860 he built a small sugar mill powered by mules that eventually became known as the Pioneer Mill Company.

Over the next several decades the company grew and changed ownership several times. By the turn of the century Pioneer Mill Company had over 12,000 acres planted in sugarcane. In 1931 Pioneer Mill Company bought the Olowalu Sugar Company, its last land acquisition, adding 1,178 acres to its plantation for a total of 14,000 acres in sugarcane and a total plantation population of 5,000. In 1999 Pioneer Mill closed, leaving only its landmark smokestack as a reminder of Lahaina's plantation days.[178]

LAHAINA COURTHOUSE

Constructed in 1859 from the coral stones of Hale Pīʻula, destroyed by the Kauaʻula Wind in 1858, the Courthouse was the center of new government services; a custom-house, a post office, a courtroom and a police court, and offices for the Governor of Maui, the District Attorney, and the Sheriff.

Now under the management of the Lahaina Restoration Foundation, its interior is leased to non-profit organizations.[179]

THE HOUSE OF KAMEHAMEHA 1853

Princess Victoria Kamāmalu, Prince Alexander Liholiho, King Kauikeaouli Kamehameha III, Prince Lot Kamehameha, and Queen Kalama. Hugo Stangenwald. Bishop Museum

Lili'uokalani, c. 1908. Photo by Harris & Ewing. Library of Congress, Prints & Photographs Division, LC-H25-13936-MM [P&P]

The Kamehameha Dynasty would remain in power until the death of Kamehameha V in 1872. Dying without naming an heir, the next monarch of the Kingdom was the elected King William Lunaliho, followed by King David Kalākaua, whose sister, Queen Lili'uokalani, would be Hawai'i's last monarch. Illegally overthrown in 1893 by American sugar planters and businessmen, Her Majesty died in 1917.

PIONEER INN

By the end of the 1800s, many inter-island steamships were calling at Lahaina and there was a need for overnight accommodations for travelers. Built in 1901, this turn of the century building was the only visitor accommodation in West Maui until the late 1950s. Maintaining its historic character, the Inn expanded in 1965 to meet the new demands of a growing visitor industry.[180]

WO HING TEMPLE

Chinese contract laborers were the earliest immigrants for plantation work. Many opened businesses after their labor contracts expired and became an economic force in building Lahaina. In 1909 the Chinese in Lahaina formed the Wo Hing Society and then built this temple on Front Street in 1912, using private donations. It served as a meeting hall and a place for religious ceremonies. As the Chinese population of Lahaina declined, the temple fell into disrepair.

In 1983 the Lahaina Restoration Foundation signed a long-term agreement with the Wo Hing Society to repair the Temple and operate it as a museum.[181]

Cookhouse of Wo Hing Temple

HONGWANJI MISSION

On Waine'e Street, next to Wailoa Church, the Lahaina Hongwanji Mission was built in 1910 on land that was once the homesite of Governor Hoapili.[182]

Many Buddhist sects came to Hawai'i in the late 1800s to serve the religious needs of Japanese immigrants. The two most predominant sects were the Hongwanji and the Shingon, the denominations of Hiroshima, Kyushu, and Yamaguchi, where immigration centers offered laborers three-year contracts.

SHINGON MISSION

The first temple of Shingon Buddhism in Hawai'i was established in Lahaina in 1902 by the Reverend Hogen Yujiri.[183] Both missions have actively served the community for over 100 years.

JODO MISSION, MĀLA

After the passage of the Reciprocity Treaty with the United States in 1876, the growth of the sugar industry required more imported labor.

In 1885 Japanese contract laborers began arriving and by 1900 over 61,000 Japanese were living in Hawai'i. By 1910 an additional 110,000 had arrived and by 1920, 42 percent of the islands population were Japanese.[184]

In 1894 Jodo Buddhism was introduced to Hawai'i and in 1912 the Lahaina Jodo Mission was founded through the support of the Japanese laborers from the many nearby plantation camps. Located on Pu'u Noa Point near Māla Wharf, the original temple was destroyed by fire in 1968 and rebuilt on the same spot in 1970. Constructed by traditional Japanese craftsmen, the roofs of both the Temple and Pagoda are made of interlocking copper shingles secured without nails. The Pagoda is 90-feet tall and holds the urns of deceased members.

Pagoda, Jodo Mission

THE GREAT BUDDHA

The Great Buddha and the Temple Bell were completed in 1968 to commemorate the Centennial Anniversary of the first Japanese immigrants to arrive in Hawai'i. The Great Buddha is 12-feet tall and made of copper and bronze. Cast in Kyoto, Japan, it weighs over 3½ tons and is the largest Amida Buddha outside of Japan. The beautiful temple grounds are dedicated to the deceased ancestors of the mission. The Temple Bell is rung eleven times at 8 o'clock each night.[185]

Temple Bell. Photo byVacclav at Dreamstime.com

PUʻU PIHA CEMETARY

Puʻu Piha Cemetery at Māla

Built upon a large sand dune, Puʻu Piha Cemetary holds the early burials of Hawaiians, as well as Chinese and Japanese plantation workers.

Once the Kapaʻula Stream (now called Kahoma Stream) ran under the Moʻaliʻi Bridge and into the Alamihi fishpond that was once below Puʻu Piha. Two small *heiau* stood here named Waiʻie and Lua kona. This whole area was known as Ka-puna-kea Village and included the massive Heiau Halulu-koʻakoʻa. All three *heiau* have been destroyed. However, for one, the story lives on.[186]

HEIAU HALULUKOʻAKOʻA

Just mauka of Māla Wharf stood the ancient village of Ka-puna-kea (The White Spring), named for the fresh water spring of that name that bubbles up on the shoreline at low tide. James Kahahane showed it to Mrs. Ashdown in the 1950s.[187]

Built by the ancient Hua Family of rulers, the Heiau Halulu-koʻakoʻa and its complex of smaller temples was a massive coral structure. Spreading from the beach to above the present Wahikuli housing and all the way through Kapunakea, this was an important *heiau*. Originally a temple of worship and for learning, it became a war temple to Ku-ke-olo-e-wa, the war god of Maui, where human sacrifices were made.[188]

THE LEGENDS OF MANU A KEPA

Once a beautiful chiefess named Manu-a-kepa, or Bird which Snatches, had the power to take the form of a *pueo*, or owl. She lived at Ke Ana Pueo, the Owl Cave by Hahakea Stream in Ka'anāpali. After Heiau Haluluko'ako'a had been rededicated as a war temple, she would wait until the temple priests were asleep at night and fly on silent wings to release the sacrificial prisoners and lead them silently to the cave. From Ke Ana Pueo they escaped through a lava tube to a cave under Pu'u Keka'a called Ke Ana Moemoe (The Cave of Moemoe) and could be safely in Kahakuloa by dawn. To protect her identity, the people she saved called the chiefess Wahine Pe'e, Hiding Woman. One night a temple priest saw her transform from an owl into a woman and chased her down to kill her. Manuakepa transformed herself into a conical stone near Ke Ana Moemoe and became known as Pōhaku Pe'e.[189] The beach from Hahakea Stream to Pu'u Keka'a was named Wahine Pe'e Beach in her honor, and is now known as Ka'anāpali Beach.

Pueo Kahi, the First Owl, mated with Pueo Nui Ākea, the Pure-minded Owl, who mothered all the sacred owl-birds of Maui. Pueo Kahi and Pueo Nui Ākea lived at Hahakea, where once a stream flowed from the Pana'ewa Forest to the sea at Ka'anāpali.

There are many stories told of the Sacred Owls of Maui and many versions of the rescues performed at Haluluko'ako'a.[190] In his 1931 survey of Heiau Haluluko'ako'a (site no. 11), archeologist Winslow Walker includes a story from a Hawaiian language newspaper translated by Thomas Maunupau in 1922. It tells a different version of the same story with the owl-god assisting a victim of the sacrificial stone temple, but the heroine of the story is Nailima, and there is no attachment to Wahine Pe'e. This same story was printed in the *Honolulu Advertiser* in 1934, and Mrs. Ashdown repeated it in her 1947 book, *Stories of Old Lahaina.*

However, Aunty Kai (Mrs. Alice Ka'ae) related another version of the Manuakepa story:

Once, in the very ancient times, there lived a beloved chiefess named Manuakepa. She was of a noble clan and when necessary, she could take the form of the *pueo*.

One day, the villagers of Kapunakea (The White Spring) were gathered at the temple of Haluluko'ako'a when they saw strange canoes approaching. Preparing to greet their guests, the villagers were instead attacked by the strangers who carried the Kū war god of Hawai'i. Fleeing in all directions, the villagers were not present to see the foreign priest rededicate their temple to his war god. The first slain (*lehua*) was laid at the altar, and after rituals were performed, the strangers ravaged the village of Kapunakea for food and spring water. As they slept, Manuakepa and her pet owl returned to the village, silently released the sacrificial victims and hid them in the secret Cave of MoeMoe at the base of Pu'u Keka'a.

Once her people were safe, the beautiful chiefess Manuakepa returned to her village covered in soft, white *tapa* and wearing a golden *lei* of *'īlima* blossoms on her head. The invaders looked at her in astonishment, but when she demanded that they leave, one of the chiefs laughed and told her he would take her as his prize.

Manuakepa raised her arms and began to chant. Her words echoed across the reef and up the silent mountainside like thunder. As she chanted, owls began appearing from all directions, attacking the invaders with claws and wings. The chief and his priest escaped in a canoe, but beyond the reef the canoe sank.

Manuakepa had the conch shell trumpeter recall the villagers. They cleared the village, buried the dead, and soon no sign of the invasion remained. From that time on, the name Haluluko'ako'a meant the Arrogant-roaring of Halulu (The Man-Eating Bird).[191] Halulu was the great white bird of Kānehekili, God of Thunder. An image of Halulu was carried at the top of a long pole during special religious seasons. *Ko'ako'a* can mean either fearless or coral.[192]

In 1802, after Kamehameha I designated his son Liholiho as heir, Haluluko'ako'a was one of the temples the five year-old Prince helped to rededicate to Kūka'ilimoku, the war god of Hawai'i. Later, after Kamehameha died, Hoapili Wahine had hundreds of coconut trees planted on the *heiau* in repentance of past sacrifices and as a symbol of prosperity.

By 1931, only a few fragments of walls remained. Most of the stones had been removed to use as ballast on the sugarcane railroad.

At one time, members of the Maui Historical Society maintained the "Royal Coconut Grove" in honor of Hoapili Wahine (Kahiehiemālie).[193]

Remains of Lua-pau (pit used for the bones of victims) of Heiau Haluluko'ako'a, Kapunakea, 1938

MĀLA WHARF

Old Māla Wharf

The flourishing sugar industry caused an increased need for shipping facilities that allowed for on-loading of cargo. Against the warnings of many *kama'āina* (native born) who advised against building a wharf near Pu'unoa Point where the Twin Currents can be destructive, Māla Wharf was built in 1922 at a cost of $250,000. Considered a modern masterpiece at the time, on opening day the first ship to tie up was damaged by rough sea. Thereafter, vessels stood outside and passengers were transported from the wharf in small motor launches.[194]

Māla Boat Launch

KAʻANĀPALI

In ancient times Kaʻanāpali was the name of the entire district from Honokōwai to Kahakuloa, across the many hills and valleys of Mauna Kahālāwai. The 1885 map of Maui by the government surveyor, W. D. Alexander, shows "K a ʻ a n ā p a l i" stretching across the upper half of Maui's head. Literally, the name means "The Cliff Caves" in reference to the many caves in the uplands of Mauna Kahālāwai. The *hūnā* and/or *kaona* meaning of Kaʻanāpali is "To Share the Heights," as translated by the Reverend Edward Kapoʻo.[195]

KE ʻAWAʻAWA

Royal Kaʻanāpali Golf Course, South

On the golf course below the first overpass bridge as you enter Kaʻanāpali Resort, is the area once called Ke Awāwa (The Valley). Bordered by Hahakea Stream that once flowed to the ocean here, this was the site of the bloody battle between the sons of Maui King Kekaulike back in 1738. Kamehameha-nui had inherited the Maui realm from his father Kekaulike, because he was also the favored nephew of King Alapaʻinui of Hawaiʻi and Kekaulike wanted peace between them. Kauhiʻaimokuʻakama was the older son and Kekaulike's leading warrior and could not accept this loss. With the aid of Chief Pele-ʻio-hō-lani of Oʻahu, Prince Kauhi challenged his brother in a tragic battle that raged for two days. So many *aliʻi* lost their lives that the battle became known as Koko-o-nā-moku, or Blood of the Islands. After the battle, the area changed from Ke Awāwa (The Valley) and became known as Keʻawaʻawa, ʻawaʻawa meaning tragic battle.

THE BATTLE OF KOKO O NĀ MOKU

Kauhi's Last Stand. Painting by Herb Kane

BLOOD OF THE ISLANDS

Kamehameha-nui, with the help of his uncle King Alapaʻinui of Hawaiʻi, and his heir and son Keawe-o-pa-la, regained rule of the Maui realm. Kauhi was taken prisoner and was sacrificed in punishment, at the Heiau of Halulu-koʻakoʻa.[196] Other stories say Kauhi was burned at Heiau Waʻilehua (Broken Warrior) at Mākila Point.[197] The first killed in battle was called a *lehua*, a valuable prize to be offered at the altar of Kū.

PUʻU KEKAʻA

Photo courtesy of Sheraton Maui Resort & Spa

Long before Lahaina became the center of power, Kekaʻa, in the Kaʻanāpali district, was the capitol of Maui's earliest family of rulers, the Hua Dynasty.[198] Said to have come from the *hema* (south), the *Mōʻi* (Kings) of Maui descend directly from Haho, son of Paumakua, ancestors of the Hua. From as far back as the 10th century, the Hua Family ruled all of Lahaʻāinaloa (West Maui) and had headquarters at Honokōhua, Honokōwai, and Kekaʻa in Kaʻanāpali. By 1350 Mōʻi Kamalo-o-Hua ruled all of Maui from Kekaʻa, which included the hill and promontory of Puʻu Kekaʻa and surrounding areas now known as Kaʻanāpali. Most of the earliest *heiau* on Maui are attributed to the Hua Dynasty.[199]

Puʻu Kekaʻa (now called Black Rock) was the sacred "Hill Holding Back the Sea," in reference to Pele's first home on Maui. Figuratively, it meant "Turning Point Hill," in reference to its use as the ʻUhane Lele, where souls leapt into eternity with the setting sun.[200] To the north of Puʻu Kekaʻa was once a royal fishpond called Loko iʻa Kapu-kapu, watered by the Honokōwai Stream.[201]

Three generations after Kamalo o Hua, Kaulahea the Great ruled Maui and Oʻahu from Lele. As a warrior prince he had conquered the Kingdom of Oʻahu for the Maui realm. His sons, Kakaʻe and Ke-ka-ala-neo inherited the realm and ruled jointly from Kekaʻa around 1450 AD.

King Kakaʻe married his maternal aunt Kapo-Hauola and they had Kahekili-nui-ʻahu-manu I.[202] His grandson would be Piʻilani, the great patriarch of the Piʻilani Dynasty.

King Kaka'e eventually followed his Priestly Order of Kāne and retired to a cave in 'Īao Valley, where he taught the *Kānāwai Akua* (Laws of God). King Kaka'e designated 'Īao Valley as the sacred burial place of only the worthy *ali'i*.[203]

King Ke-ka-ala-neo ruled from Keka'a and had command of all Maui. He is credited with introducing the feather cloak, *'ahu'ula*, as royal attire of the *ali'i-nui* (high chiefs).

THE LEGEND OF 'ELE'IO AND THE FIRST FEATHER CLOAK

Warrior, Sandwich Islands. Hand colored by Nicholas-Eustache Maurin, lithograph by Jacques Arago. Honolulu Museum of Art

The ruling chiefs kept *kūkini*, swift runners who could cross the island with great speed and deliver messages or obtain goods for their chief. An *'elele* was another class of runner. They were chiefs and were especially fast. They were like an ambassador or diplomat. Even though *kūkini* were most often commoners, they were honored for their loyalty and trustworthiness. *Kū* means standing, upright. *Kini* means populace, multitude. They stood above the populace and were highly regarded and valued.[204]

'Ele'io was a chief of Hāna and a famed *'elele* for King Keka'alaneo. He could circle the island three times in a day and still not be tired. He ran so fast that fish from the ponds at Hāna were still alive when he arrived at Keka'a.

One day King Keka'alaneo sent 'Ele'io to Hāna to fetch *'awa* (drink made from kava root) for his supper. Chief 'Ele'io was also a *kahuna* (priest) with the ability to see spirits, and on his journey he met the spirit of a lovely chiefess named Kanikani'ula. So beautiful was Kanikani'ula that 'Ele'io decided to try to restore her spirit to life. He built a bower of perfumed vines and invoked the assistance of the gods as he performed the *Kapukū*, a ceremony to restore life to the dead.[205]

'Ele'io was successful, and in gratitude Kanikani'ula presented 'Ele'io with a finely woven cloak adorned with thousands of tiny brightly colored feathers. 'Ele'io had never seen such a glorious garment or such fine craftsmanship. Kanikani'ula explained that she had learned the secret of featherwork from her spirit ancestor when she was in spirit form.[206]

By now 'Ele'io was very late returning, and knew death would be the punishment. But he returned to the King's court with Kanikani'ula, the feather cloak, and the King's *'awa*. When 'Ele'io presented the feather cloak to Keka'alaneo, the King and his court were amazed by the glorious garment. 'Ele'io was restored to favor. So beautiful was Kanikani'ula that King Keka'alaneo asked her to be his Queen.

The feather cloak became known as 'Ahu o Keka'alaneo (Cloak of Keka'alaneo) and was eventually passed down the generations to Chief Kanaina, the father of King Lunalilo (c. 1874). After his death in 1878, it was purchased by the government for $1,200 and was listed in the inventory of Bishop Museum in 1899 by its director, William Brigham.[207]

King Keka'alaneo married Kanikani'ula of the Kama-ua-ua family of Moloka'i. Their son and heir was Ka'ululā'au, about whom many legends are told.[208] Kapō of the Deep Sea had gifted Keka'alaneo with breadfruit seedlings which the King had planted all over Lele, providing a wonderful food source and gracious shade.

THE LEGEND OF KA'ULULĀ'AU

Hawaiian Breadfruit, 1890.
Painting by Persis Goodale Thurston Taylor

Prince Ka'ululā'au was a mischievious young man and kept pulling up breadfruit seedlings despite his father's admonishments. Finally he was banished to Lāna'i, known then as the Island of Demons, where evil spirits had overtaken the island and the people were ill and dying. There was little hope for the young prince when an aged *kāhuna* (priest) took pity on him and gave him the sacred Spear of Lono.

With this powerful spear Ka'ululā'au destroyed the evil spirits and restored Lāna'i and her people. When his father saw Ka'ululā'au's signal light, he knew he had survived and sent a canoe to fetch him home to Lele to take his rightful place as Prince Ka'ululā'au (The Breadfruit Grove).[209]

'UHANE LELE

Pu'u Keka'a was the 'Uhane Lele, or "Soul Leaping Place" of West Maui. Every island had an 'Uhane Lele at its most western hill where souls leapt into eternity. The *'uhane* (soul) *lele* (travels) into Kahiki Moe (the Horizon where the sun sets) and enters Nā Moku Hūnā o Kāne, (the Hidden Isles of Kāne, the Creator, land of eternity and new life). Each soul, each spirit, is of Kāne, lives in Kāne, and returns to Kāne.[210] The belief is that the soul travels west, like the sun, and rises again, like the sun, in peace and perfection, if the dead one has observed the *Kānāwai Akua* (Laws of God).[211]

On top of Pu'u Keka'a stood a small heiau about 20 feet square with walls 3 to 4 feet high. It was called Ho'akua (*Ho* – to give, *'akua* – god).[212] Inside the walls and in alignment with an opening in the west wall stood an altar dedicated to Kanaloa. Facing 'Au'au, the Ancestral Sea or channel between Maui and Lāna'i, the officiating priest would recite the final *pule polo*, or prayers for the dead, at the setting of the sun. Pu'u Keka'a was sacred ground with its own *kapu*, and the people remained at the base of the hill to listen and pray.[213]

Pu'u Keka'a from Ka'anāpali Beach

KAHEKILI LEAPS

Warrior Cliff Spirits. Painting by Herb Kane

Three centuries after Kaʻululāʻau, the last great King of Maui, King Kahekili-nui-ʻahumanu II, The Thunderer (c. 1765), makes Puʻu Kekaʻa famous in legend. Kahekili was a man of perfect lineage, directly descended from Kāne, and was considered a god. A noted warrior, Kahekili enjoyed all athletic games but was a champion at Lele-kawa, leaping from high cliffs into the ocean. He was famous for jumping the sea cliffs at Kahakuloa and Hāna.

One day Kahekili visited his people at Kekaʻa and saw that they had grown lazy and had been led astray by foreign influences and drunkenness. The taro patches, potato fields, and sacred fishpond were being neglected. Kahekili reminded the people that his laws were God's law, the *Kānāwai Akua*, and they must obey. To prove his godliness to the wayward people of Kaʻanāpali, Kahekili dared to leap from sacred Puʻu Keʻkaʻa, the ʻUhane Lele where only the souls of the dead leapt into eternity. As he walked to the summit the people began wailing. They watched him pray at the altar, then run and dive into the dying sunset. When he returned unhurt, the people fell to the ground and wept in shame for doubting his divinity. Kahekili forgave them, the priest cleansed them in the ocean, and Kekaʻa prospered once again.[214]

PŌHAKU MOEMOE

Pōhaku Pe'e and Pōhaku Moemoe rest at the southern end of the Maui Eldorado on Keka'a Drive

Long ago, when god spirits walked the earth, Moemoe, the God of Sleep, lived in a cave below Pu'u Keka'a named Ke Ana Moemoe. Māui-a-Kalana lived at Kahakuloa and travelled Ke Alanui Kī-ke'eke'e o Maui, the Zigzag Trail of Maui, on his quest to slow the sun. On his way he stopped at Pu'u Keka'a where Moemoe taunted him, telling him he will never catch the sun and will die trying if he does not stop and rest. Māui promised Moemoe that he would succeed and that Moemoe would be the one to die. When Māui accomplished his heroic deed of making the days longer, he kept his promise and turned Moemoe to stone. The six-foot long Pōhaku Moemoe lay near the smaller Pōhaku Pe'e by Ke Ana Moemoe (the Cave of Moemoe) at Pu'u Keka'a. When the Ka'anāpali Resort was being developed in the early 1960s, members of the Hawaiian Civic Club were concerned for the Pōhaku and wanted them saved. Mrs. Ashdown went to Eric Jacobsen of the Maui Eldorado and asked him to save the stones, which he kindly did. They now rest together on the edge of the Royal Lahaina golf course and are marked by a Ka'anāpali Historical Trail sign.[215]

MADAM PELE'S FIRST HOME

Pu'u Keka'a is said to have been Pele's first home on Maui. Her sister, Kai-po (or Kapō) of the Deep Sea, chased her on to Pu'u La'ina where she rested until the water spirits made a lake for the home of the infant *Mo'o* of the Kiha Family. Pele took her torch and moved on to build 'E'eke with its flat top. Soon the rain spirits came to visit with the clouds and the *pili* grasses became sodden. Pele fled to Pu'u Kukui where she was drenched until the pond called Ki-'o-wai-o-Kiha (now known as Violet Lake) was formed.

Angered by these intrusions on her powerful fire, Pele raced across the valley of mirages and went straight up to the top of the East Maui mountain, in the pathway of the sun. Here she dug a huge home and when the people saw the fires they named the mountain as Mauna Ka'uiki, or Mountain of First Fires Glimmering, which today is known as Hale-a-ka-lā, House of the Sun.[216] Pele found her permanent home in Hale Ma'uma'u on the island of Hawai'i, where her fiery torch still burns.[217]

NĀ HONO A PI'ILANI

The Bays of Pi'ilani

Once Prince Ka'ululā'au (c. 1475) returned from Lāna'i and had grown into a young man worthy of his title, King Keka'alaneo went in search of a special wife to be the mother of his son's heirs. Keka'alaneo voyaged on a rainbow to La-keo-nui-a-Kāne, (Land mass belonging to Kāne), where he met a beautiful maiden whose name was Pi'ilani (*Pi'i* – to rise, ascend; *lani* – heaven, spiritual). She fulfilled every ideal he had hoped to find for his son and she was intrigued by the many desirable qualities the King told her his son possessed. Pi'ilani sent her spirit on a red cloud to look for Ka'ulula'au and the cloud settled on Mauna Lei, the mountain of Lāna'i. The people of Lele saw the cloud and knew what it meant. So did Prince Ka'ululā'au. He went into a trance-like sleep and sent his spirit to meet the spirit of Pi'ilani on Mauna Lei. The two spirits met and fell madly in love. Pi'ilani's spirit flew to Lele where she took her human form and awakened the sleeping Prince Ka'ululā'au. They married and lived together at Lua'ehu when he became King of Maui. In her honor King Ka'ululā'au named the six fertile valleys and *hono* (bays) of his homeland for the beauty of their love and marriage: Hono-kōwai, Hono-keana, Hono-kōhua, Hono-lua, Hono-nana, and Hono-kōhau.

It was said that the *hūnā* (hidden) and the *kaona* (poetic) translations of the names describe the various stages of love making and conception.[218]

These fertile lands and bays were the pride and joy of all the *ali'i*. Two generations later, Ka'ululā'au's ancestor Pi'ilani (c. 1525) becomes Maui's great king and Nā Hono-a-Pi'ilani, The Bays of Pi'ilani, eventually become associated with King Pi'ilani, but the honor lies with the beautiful wife of King Ka'ululā'au, Queen Pi'ilani.

Several other versions of this story were told by West Maui Hawaiian Civic Club members in the 1930s, with Kama or Kamalala-walu (c. 1575) as the husband, but they all said Nā Hono a Pi'ilani was named for the beautiful wife. These stories of Nā Hono a Pi'ilani and translations were told by Aunty Kai (Mrs. Alice Ka'ehukai Shaw Ka'ae), Lahela Reimann, and Mary Chan Wa.[219]

HONOKŌWAI VALLEY

Bay and Valley of Waters

Honokōwai Valley from Pu'u Kukui. Photo courtesy of Karl Magnacca

From Pu'u Kukui and Lake Manowai come the headwaters of Honokōwai Valley that in ancient times flowed into Loko i'a Kapu-kapu (very restricted), the Royal Fishpond at Keka'a. Kō-wai (*kō* – fulfilled; *wai* – water) signifies this valley as very sacred and the *kapu* was lifted only on special occasions.

Lake Manowai (Source of Life) feeds the First Springs of Kāne and Kānaloa through its outlet 'Ōmaka, The Beginning, into the headwaters of Honokōwai Stream that flows into Honokōwai Bay, The Bay of Waters.

Honokōwai Valley has an *ana-pe'e-kaua* (war-hiding cave) where the *ali'i* Kaulahea (c. 1675) escaped with his people. It was said that he was led to the vine-covered entrance by his *'aumakua* (guardian spirit). The cave is called Piliwale (to bring together) and was used during times of warfare and heavy storms.[220]

Along the north side of Honokōwai Valley is the Haelā'au Trail that goes up to Pu'u Kukui where rain gauges track the source of West Maui's water. Along the trail is an area called Nākulalua, where medicinal priests lived and grew herbs. They flew a pennant for people seeking them out and the area became known as Haelā'au (Leaf Flag). Below the Haelā'au area is the land called Mo'omuku (Whispering Spirits), between Honokōwai and Mahinahina. Older Hawaiians told Mrs. Ashdown that some claimed the first ancestors settled here and the spirits of those early people still live here as *'aumakua* to watch over their descendants.[221]

THE LEGEND OF PŌHAKU KA'ANĀPALI

Between Lae o Kama in Mahinahina and Kapua Bay of Kahana lies a very large boulder named for the noted seer of Kapua called Maika'i-hele-ana-pali. Maika'i was an *ali'i* of great *mana* (spriritual power) and a champion athlete noted for his cliff climbing. He was still strong in his old age and by then his renown as a great cliff climber had spread throughout the islands.

On the island of Moloka'i was a young chief so proud of his athletic abilities that he was known as The Boaster. He challenged Maika'i to climb the great sea cliffs of Moloka'i. Maika'i accepted the young Boaster's challenge and won the contest. Angry, The Boaster challenged Maika'i to another contest. Maika'i knew the young Boaster needed to be taught a lesson, so he invited him to come to Ka'anāpali to climb his cliff. If he failed, he would lose his life and all his possessions. The Boaster accepted.

When they beached their canoes at Kapua Bay, Maika'i pointed to a large boulder by the sea near the road and said, "Here is my cliff. See if you can climb it." The Boaster laughed at Maika'i, "You silly old man, I can climb that rock in one leap." But The Boaster could not climb it and finally fell exhausted to the ground, realizing that it was Maika'i's superior spiritual power that beat him. "I have lost. Take my possessions and my life." But Maika'i told the young chief to take his possessions home and to give thanks to the gods for them. The chief built a *heiau* in Halawa Valley and became a teacher like Maika'i-hele-ana-pali. The rock became known as Pōhaku Ka'anāpali.[222]

Pōhaku Ka'anāpali was eventually covered over when the road was widened, but the name remains on old maps and so does its inspiring story.

Kapua Bay sounds like The Flower or The Child, but the name is in reference to the spawning fish attracted to the many undersea fresh-water springs in the bay.[223]

HONOKEANA

Bay of Satisfaction

A very tiny bay south of Nāpili Bay, this *Hono* is unlike the five other *Hono* and doesn't have a Valley or Stream attached to it. Perhaps it was chosen as the Bay of Satisfaction for its obscure privacy. Some said Honokeana was a *kaona* (poetic) reference to the bride's private parts.[224]

Honokeana carries its name in the *ahu'pua'a* (land division) of the area.

Nāpili Bay sunset. © 2016 Watersun Photography ~ watersunphotography.com

THE LEGEND OF NĀPILI BAY

Long ago, at a village not far from Honokōwai, lived a chief whose son was named 'Ehe-hene. The village *kāhuna* was of great renown and had a beautiful daughter named Hi'iaka, after her ancestress Hi'iaka-i-ka-poli-o-Pele. 'Ehe-hene and Hi'iaka were inseparable companions who enjoyed playing for hours in the surf with their friends. One day Hi'iaka swam out too far and was grabbed by the shark spirit Kauhuhu of Kīpahulu. 'Ehe-hene dove deep into the sea and was so shocked to find Hi'iaka's lifeless body that he too, drowned.

When the lifeless bodies of the two companions floated to the surface, two fishermen carried them to shore in their canoes. Hi'iaka's *kāhuna* father saw two *nai'a* (porpoise) leaping together alongside the canoes, escorting them to shore, and he knew they were the sea-bodies of the young companions' spirits, together forever. In their honor, the village was named Nāpili, The Companions.[225]

HONOKAHUA

Bay and Valley of Happy Meetings

Pu'u Kahuahua, Honokahua

From the earliest times, Honokō-a Hua was home to the Hua Dynasty, who ruled from 1015 to 1400 AD. Pu'u Ka Huahua, now known as Pineapple Hill, and the whole *ahupua'a* (land division) of Honokahua was a family base for the *Ali'i o Maui* who possessed the *kapu* of the Burning Sun.[226] Recent studies of ancient burials have shown a continuous settlement of the area from 600 AD to 1800 AD, with many high ranking *ali'i*.[227]

THE BATTLE OF KAHUAHUA

Because two brothers of the Hua Dynasty fought a bloody battle against each other here, the hill became known as Pu'u Ka-Hua-Hua, Brother against Brother. The battle took place from Pu'u Kahuahua down to the beach of Honokahua Bay and was known as the Battle of Kahuahua.[228]

Mrs. Ashdown and her family lived at Honokahua for nearly thirty years and often saw exposed *iwi* (bones) when the winds shifted the sand dunes of Honokahua.

PUʻU KAHUAHUA

Pineapple Hill Road, Kapalua Resort

At the top of Pineapple Hill Road, lined with Cook Pines planted over a hundred years ago, sits Puʻu Kahuahua, once a headquarters and the royal playground of the Hua Dynasty. Sporting events were a serious pastime among the chiefs, and this hill was the center of activity with *kahua holu* (grass sledding course or slide) for the games of *ʻulu-maika*, which resembled bowling, and sledding on their *papa holu* (sleds). Here the Hua chiefs trained and competed in *lua* (martial art), wrestling, spear hurling, and games such as *kōnane* (a type of checkers played with pebbles).

When thought of as the playground or sports center, the area was referred to as Hono-kahua (*kahua* – open space for sports) and said as Hono-kōhua (Bay of Hua) in reference to the kingly family.[229]

D. T. Fleming, Manager of Honolua Ranch, built his stone house on this hill in 1915 and named it Maka-ʻoi-ʻoi, "sharp eyes" (or clear view). The home eventually became the landmark Pineapple Hill Restaurant when Honolua Ranch became Kapalua Resort in the 1970s.[230] The restaurant eventually closed.

JAMES YOUNG KĀNEHOA

James Young Kānehoa.
Bishop Museum

Liholiho Kamehameha II inherited the *ahupua'a* of Honolua and Honokahua from Kamehameha I, who won these *ali'i* lands when he conquered Maui in 1790. James Young, second son of John Young, then chiefly advisor to Kamehameha I, was born the same year as Liholiho and they were close childhood friends. Liholiho Kamehameha II named him Kāne Hoa, Brother-Companion, and he became officially known as James Young Kānehoa. In November of 1823, Kānehoa accompanied King Kamehameha II and his Royal Party to England as interpreter for the King. Sadly, in July of 1824, King Liholiho and Queen Kamāmalu succumbed to the measles while waiting to be presented at the Court of St. James. Along with Chief Bōki and Chiefess Liliha, Kānehoa escorted the bodies of their Royal Majesties home to Lahaina aboard the *HMS Blonde,* provided by the British Government and arriving in Lahaina in May, 1825.

Kamehameha III inherited the Kingdom from his brother at the age of 11 years, and this included the lands of Honolua and Honokahua.[231]

Kānehoa married Sarah Davis, daughter of Isaac Davis, Kamehameha the Great's chief counselor and gunner. They had no children but adopted the son of his sister Jane Lahilahi and her husband Joshua Kaeo.

Working in partnership with Kamehameha III, Kānehoa planted the Mahana Coffee Plantation above Honokahua Bay and years earlier had planted sugar on lands in Honua'ula that eventually became part of 'Ulupalakua Ranch.[232]

Kānehoa had a stone house by the Hawaiian church at Honokahua Bay and the road that went upland to Mahana Coffee Ranch and the forests above.

In 1845, at the age of 48, Kānehoa was appointed to the House of Nobles. In 1846 Kamehameha III appointed him Governor of Maui, Molokaʻi, and Lānaʻi, as well as a member of the Privy Council where he served His Majesty in all capacities until his death on October 1, 1851.[233]

In 1852 W. D. Alexander surveyed all of Maui for the Kingdom and it was revealed that Kānehoa owned the land of Mahana in Honokahua and the land of Paeʻahu in ʻUlupalakua.[234]

Kānehoa was with his sister Grace and her husband Dr. Rooke at their home in Honolulu when he died. At his request, he was brought back to Maui and buried below Mahana Coffee Plantation, next to his adopted son Keliʻiaikai Kaeo, who died in infancy.[235]

Honokahua Bay became a county beach park in 1975 and was named D. T. Fleming Park, for the late David T. Fleming, manager of Honolua Ranch. Across the street from the park is the old Kalawina (Calvinist) Church, now a preschool. On the knoll just above it is a stone wall encircling the grave of Kānehoa and other members of the Young and Davis families.[236]

Kānehoa's niece Emma married Kamehameha IV and became Queen Emma.

Old Hawaiian Calvinist Church, Honokahua Bay

HONOLUA VALLEY

Bay and Valley of Twin Seeds

Honolua Bay, Honolua. Photo by John DeMello

HUA-KAʻI-PŌ

Honolua Valley has always been the home of powerful spirits and was known particularly as a place of the Hua-kaʻi-pō, the Night Marchers, or the ʻŌi-o, Spirit Travelers. People would not travel there on certain nights for fear of meeting the spirits of dead *aliʻi* walking the old trails. Unsuspecting travelers would have to quickly prostrate themselves to avoid the spears carried by the Night Marchers.[237]

The Royals could travel on the moon nights of Akua, Kāne, and Kanaloa, but these nights were *kapu* to any others. The true night of Akua does not come each month but on any early morning when the waning moon is a thin sliver and a star is in the horn of the moon. According to Hattie Kanaka-o-kai Yoshikawa, this is really when the spirits walked, on this very sacred moon night when drums and chanting could be heard.[238]

These bays and valleys are the last undeveloped lands in West Maui that still contain the ancient stones of *heiau* put in place centuries ago, most likely by rulers of the Hua Dynasty. Though the purpose and translations, for the most part, have been lost to time, the names remain with the stones as tributes to an ancient past.

HONOLUA RANCH

Honolua Valley. Photo by John Carty

Honolua Ranch was begun in the lush Honolua Valley by Richard Searle, an ex-sea captain from England who settled in the valley with his beautiful Hawaiian Chiefess wife who had inherited lands from ancestors of the Kamehameha, Konia, Lunalilo, Davis, and Young families. The ranch supported several Hawaiian and part Hawaiian families in the "old style" by raising cattle, horses, coffee, and taro, as well as fishing.[239]

In 1892 H. P. Baldwin acquired the land and invested in the ranch. Under the management of Richard Searle, Honolua Ranch soon had a poi shop and butcher shop in Lahaina and another poi shop in Wailuku. The ranch shipped cattle to Honolulu and Moloka'i from the Mahinahina-Honokōwai area.

In 1902 Honolua Ranch and Pioneer Mill Company partnered to build the Honokahua Ditch, harnessing much needed water for agriculture. Two years in the making, it was completed in 1904.[240]

After the death of H. P. Baldwin in 1911, D. T. Fleming was hired to manage the ranch.[241]

DAVID THOMAS FLEMING

David T. Fleming came to Hawai'i from Scotland as a child in 1889. His father, an agriculturist, had been hired by Henry Perrine Baldwin to manage Grove Ranch in Ha'ikū. Schooled at home under the tutorship of Miss Greene, he was an avid learner and spoke fluent Hawaiian. Under the tutelage of H. P. Baldwin and his son Dr. Will D. Baldwin, an authority on agriculture and particularly on a new product, the pineapple, young Fleming gave himself enthusiastically to the experimentation of this new fruit.[242]

Fleming went to college on the mainland and majored in Hydrology and Watershed Management and returned home in 1903 to become time-keeper on the Hāmākua Ditch project in Ha'ikū.[243]

In 1911 Fleming accepted the offer to manage Honolua Ranch on the condition that he would have free rein to turn the land into a pineapple plantation. He began by the building of the Honolua Ditch to replace the leaking Honokahua Ditch and started planting pineapples, first just four acres, and soon all the seedlings that could be supplied from Dr. Will Baldwin's Ha'ikū experimentation farm were being planted. By 1914 Honolua Ranch was shipping 6,000 cases of pineapple a week to the mainland.[244]

In 1915 the Honolua Ditch was completed and Mr. Fleming moved the Honolua Ranch headquarters to Honokahua, where he had built his new home on top of Pu'u Kahuahua. Here he established a cannery, company offices, and a company store.[245] Although the Honolua Ranch was now headquartered in Honokahua, it retained the Honolua name.

Eventually a larger cannery was built near Māla Wharf. Pineapples were brought in by train from Honolua Ranch, processed, and then shipped out from Māla Wharf or the Ka'anāpali Wharf. Now known as Baldwin Packers, Ltd., by the 1930s there were over 9,000 acres planted in pineapple.[246]

While building the pineapple plantation, Fleming was also dedicated to reforesting the watershed above Honolua which had been stripped long ago of *koa* and *'ōhi'a* trees to pay taxes. The stately rows of eucalyptus and Norfolk Island pine trees that now grace the grounds of Kapalua Resort were planted by D. T. Fleming 100 years ago as part of his reforestation program. He planted a 75-acre arboretum of teak and other hard woods in Honokahua in hopes of one day developing a lumber business. Now called Maunalei Arboretum, it is now part of the 22,000 acre Kapalua Resort and can be visited by resort guests.[247]

D. T. Fleming was actively involve in Maui's civic affairs and served as President of the first Board of Water Supply.

After extensive travel searching for new and exotic plants to help diversify Maui agriculture, Fleming devoted his retirement to the preservation of Maui's endangered endemic plants. As a reward for solving the infestation of a poisonous plant in the pastures of 'Ulupalakua Ranch, Edward Baldwin gave Fleming his choice of land for a native plant arboretum. He chose Pu'u Māhoe, 17 acres at Auwahi, a native dryland forest above Keonioio on Haleakalā's southern flank. The D. T. Fleming Arboretum at Pu'u Mahoe is today Hawai'i's oldest and largest native arboretum working to preserve dryland species and restore Maui's leeward watershed. [248]

D. T. Fleming died in 1955. In 1975, Maui County dedicated the Honokahua beach park as D. T. Fleming Beach Park in his honor.

In 1969 the land holdings of Honolua Ranch became part of Maui Land and

D. T. Fleming Beach, Honokahua Bay

Kapalua Resort lands. Photo by Forest and Kim Starr

Pineapple Company, a new entity created by separating the holdings of a branch of the Baldwin Family. Colin Campbell Cameron, great grandson of H. P. Baldwin, became President and CEO, and in 1975 incorporated Kapalua Land Company as a subsidiary. Through this company Colin Cameron created the elegant Kapalua Resort community comprised of 22,000 acres of former Honolua Ranch land.

After 97 years, once the largest producer of private label pineapple in the nation and the largest employer on the island of Maui, the company closed down all pineapple operations on December 31, 2009.

Kapalua Beach, Kapalua Bay

HONOKAHUA BURIALS

Honokahua burial site 1

In August 1987, Kapalua Land Company began the excavation of an ancient sand dune above Honokahua Bay to make room for the new, fully-permitted, Ritz-Carlton Hotel. A team of archeologists began the meticulous disinterment of the known burial site. Sixteen months later, after the media revealed that the excavation had exposed the *iwi* (skeletal remains) of 900 pre-contact native Hawaiians, with more to come, the Hawaiian community became outraged statewide. They mobilized in protest against the ongoing disinterment and demanded an immediate halt to the excavation.

A major controversy followed and in an unprecedented move Governor John Waihe'e, the state's first native Hawaiian governor, interceded and asked for a re-evaluation of the disinterment of the remains at Honokahua, calling the site "one of major historical significance" and announcing that he had appointed his directors of the Office of State Planning and the Department of Land and Natural Resources as a state task force to work with the various parties to find a solution. As a result, these agencies, the mayor, the landowner, the state Office of Hawaiian Affairs, and several native Hawaiian groups led by *Hui Alanui O Mākena*, came to an agreement that the exhumation would halt, the hotel site would be moved 500 feet *mauka* (inland), the disinterred bodies would be reinterred exactly where they were found, and the State of Hawai'i would pay the landowner $5 million to place the 13.5 acre parcel in a perpetual conservation easement.[250] As Governor Waihe'e said, "It [is] just not a place to build a resort hotel. Period."[251]

Honokahua burial site 2

Studies completed on the site revealed a continuous stable community dating from circa 610 AD to 1800 AD.[252] The burial site potentially holds over 2,000 individual remains. Sixteen *lei niho palaoa*, a traditional symbol of high ranking *ali'i*, were found with the remains of those removed. This is an unusually high percentage of the noble *lei* to be found at a burial site.[253] Most likely many belonged to the *ali'i nui* of the ancient Hua Dynasty.

From this tragic conflict new legislation was passed to give unmarked burial sites the same protection as modern cemeteries. Hawai'i's Burial Sites Program was created to oversee implementation of the law. Despite these safeguards, Hawaiians are still fighting to protect the *iwi* of their ancestors found "inadvertently" at development sites.

IWI KŪPUNA

Hawaiians believe that the *mana* (divine power or spiritual essence) of an ancestor resides in the *iwi* (bones) of the departed. The *mana* imbues the land where the *iwi kūpuna* (ancestral remains) lay secreted away. Many believe that moving the *iwi kūpuna* for any reason is a desecration of the *iwi*, the *mana*, and the *'āina* (land) where they were laid to rest.[254]

LEI PALAOA

Woman of the Sandwich Islands, 1819.
Louis Choris. From *Voyage pittoresque autour de monde,* F. Didot, Paris, 1822

The Lei Pa-lao-a was a sacred ornament worn only by *ali'i* of high rank. It symbolizes high birth, truth, and honor. The *ali'i* spoke as gods, with a clean tongue. They also spoke in the *hūnā* dialect so that sacred truths would not be desecrated by the uninitiated.

The tongue-shaped ivory pendant is called *elelo*, which means the tongue and speech. The *lei* is made of many narrow strands of braided hair from a revered *ali'i* ancestor of high standing. *Pa-lao-a* is whale ivory, gathered from the carcasses of dead sperm whales washed ashore, at sites that were highly guarded. Also called *lei niho palaoa, niho* means tooth. After the *palaoa* was carved with stone adze and shells, it was inserted into a stalk of sugar cane and suspended over fire to turn the ivory a rich amber color.

Nā Ko-ho-lā, The Humpback Whales, traveled with the sun and came to Hawaiian waters in the winter to birth their calves and strengthen them for the long journey back across the Pacific in the summer. *Koho* is to choose and *Lā* means the sun. The *koholā* were considered sacred and never hunted by Hawaiians.[255]

KIHA A PIʻILANI TRAIL

During a preliminary survey of the Honokahua site in 1986, a 150 meter-long section of a kerbstone trail, believed to be a portion of King Piʻilani's Highway built by his son Kiha-Piʻilani, was revealed.[256]

A stone paved trail four to five feet wide and lined on both sides by large partially buried beach stones, this *alaloa* (highway) once encircled the entire island of Maui and became known as Ke Alaloa o Maui, the Broad Highway of Maui.

King Piʻilani united the Maui Kingdom under his rule and produced a peaceful and prosperous time. He lived in Hāna and Lele (Lahaina) and built the Piʻilani Trail and stone paved *ʻauwai* (irrigation ditches) around East Maui. After his death around 1525, his youngest son Kiha-a-Piʻilani (Kiha of Piʻilani) finished the *alaloa* around West Maui and repaired and built new *ʻauwai*. Most of the Kiha-a-Piʻilani Trail of West Maui is gone now.[257]

HONOLUA BAY

Photo courtesy of Hawaii Web Group, MauiGuide.com

Along the Honolua shore once stood Puhalakau, a *heiau* for Kū'ula, the god of fishermen. Modern stone walls and houses built on the site have obliterated the structure. Further up the gulch, just east of the big bend in the road, is Heiau Honua'ula, with its remains of stone platforms and walls.[258]

North of the bay a large cliff juts out as Lae Līpoa and its broad flatland named Kula o ka'ea once had an ancient *hōlua* (sled) course that stretched across the flatland to Punalau Point and Bay. It was replaced by a stone clubhouse and a golf course named the West Maui Golf Links at Honolua. When WWII began, the course was plowed up and planted with pineapple to prevent enemy planes from landing.[259]

Honoapi'ilani Highway crosses Kula o ka'ea and makes a sharp turn mauka. Above this bend in the road sits Pōhaku Pule, the Prayer Stone, where travelers use to stop to pray for safe travel on their way to Kahakuloa.[260]

Honolua Bay is the only sheltered bay on Maui with a structural reef. Its diverse reef life is protected as a Marine Life Conservation District. With a world famous surf break, when this popular surfing and snorkeling bay was threatened with development, the citizen group SaveHonolua.org began lobbying for protection. Years of effort led to the state purchase of a conservation easement for the 720 acres of Lae Līpoa in 2014.[261]

HONOKŌHAU

Bay and Valley of Fulfillment

Photo by Makahiki87 at Dreamstime.com

At Pūnahā Gulch, just before reaching Honokōhau Valley and Bay, an ancient *heiau* sits above the roadway on the ridge at the western side of the gulch. Its name is ʻIliʻili-kea.[262] Further up the road as it curves around the east side of Honokōhau Valley, a large terraced *heiau* sits on the cliff 200 feet above the sea. Its name is Ma-iu, and was said to be once used in human sacrifice. It is one of the last *heiau* consecrated by Liholiho in 1819 before the death of his father, Kamehameha I. There is a trail leading up to the temple paved in beach stones that is said to be part of the ancient Kiha-a-piʻilani Trail.[263]

The valleys of Honokōhau, Honolua, and Honokōwai merge together at around 4,000 foot elevation, below Lake Manowai where their headwaters begin. At the apex of Honokōhau Valley is Maui's highest waterfall, dropping 1,120 feet. It appears as one majestic fall but is really the combination of an upper and lower waterfall. Named in reference to the sacred waters formed by the dieties Kāne and Kanaloa in Lake Manowai (Source of Life), the upper fall is Wē-ki-u, meaning the highest rank of their sacred waters. The lower fall is Wai-lele, Leaping Waters, in reference to the ecstasy of the Twin Waters of Life.[264] These esoteric names are forgotten now. In 1969 the state renamed the falls as Honokōhau Falls.[265]

Wēkiu and Wailele waterfalls.
Photo by Royce Bair

Taro was cultivated in *lo'i* (irrigated terraces) up these valleys, at one time thought to number 4,000. The stream water that flows through the *lo'i* in Honokōhau Valley is named Wai Ho'okō, Water of Fulfillment, where it eventually empties into Honokōhau Bay and feeds the many varieties of fish that come to spawn in the brackish waters.[266] This stream flow never ceases, even today, and at one time supported the second largest cultivation of taro, after Kahakuloa Valley, on Maui.[267]

Honokōhau stream. Photo courtesy of Harold Willome

Overgrown with time and disuse, the *lo'i kalo* (taro patches) are slowly being reclaimed and put back into production by residents of Honokōhau Valley.

Below Pu'u Kukui and Lake Manowai, where these valleys merge, is Pu'u Wai'uli (Hill of Dark Waters). Near this *pu'u* was a burial pit for the commoners of Kahakuloa to Laha'āina. During war times, people would hide the bodies of their dead in this deep pit to safeguard their bones from the enemy. The burial pit became known as Wai'uli Pit.[268]

Just north of Honokōhau Bay is a large promontory of land called Kalala'oloa, in reference to the shining greenness of the open land. A large temple of astronomy and astrology once stood here but was plowed under long ago for pineapple. This *heiau* was used for worship and education and was also a place where the *haku mele* (priestly bards and poets) composed the genealogical name-chants for the *ali'i* and high chiefs of the area.[269] Its name has been lost to time, but it is still easy to imagine a temple on the broad promontory where the priests chanted to the stars.

At Lae Kanounou, the point of this headland, a *ko'a* (shrine) once stood where secrets of the sea were taught. *Ko'a* were built to honor Kū'ulakai, who built the first fishpond and taught people to only take what was needed; to give the first fish to God at the alter of Kū'ulakai, the second to a stranger, and to share the rest with family and friends. Nounou refers to the abundance of Honokōhau Bay, once known for its *'ama'ama* (mullet) and *moi* (threadfish) so loved by the *ali'i*.[270]

Honokōhau lo'i kalo under the care of Wili Wood and Honokōhau 'ohana.
Photo courtesy of Mehana Lee

NAKALELE POINT

Nakalele Blowhole. Photo courtesy of Abhinaba Basa

On the eastern point of Kalala'oloa is Lae Nakalele, Maui's most northern point, with its namesake lighthouse, now a Coast Guard Beacon. Erected in 1908, the lighthouse was continuously manned by Hawaiian keepers until 1923, when the light was automated. West of the beacon along the shore is the Nakalele Blowhole. *Naka* means sea creature and *lele* means to burst forth.

Legend tells of a supernatural sea dragon named Nakalele that lived in a cave at Nakalele Point in ancient days and would rise from his cave, flash his rainbow colored scales, and drag unsuspecting travelers into his cave. The journey to the next valley, Kahakuloa (The Eternal Master), in those days was perilous as spirits and other supernatural beings were known to roam these lands.

Planning the journey to avoid the *kapu* nights when the Hua-ka'ipō (Night Marchers) might be out, the wise traveler started at Pōhaku Pule (Prayer Rock) with offerings to the gods and a prayer for safe travel. He then reached Pu'u Kīle'a, said to be the "basis of everything pleasant," where he could prepare himself for the rest of the journey and gather offerings he would need later. At Kalaloaloa, the winds blew wildly and scary feelings would grip the traveler. Māniania was named for these scary feelings (*mānia* – shuddering sensation.)

The traveler next came to Ke Ana-lua'hine, The Cave of the Two Women. These evil women had supernatural powers, but if the traveler left an offering of food or *'awa* (drink made from *kava* root), he could run past the women as they rushed toward the offering. At Wai-ke-akua (Water of God) was a spring made by Kāne and Kanaloa in the time of the gods. If the traveler had prepared himself properly for the trip and his *mana* was strong, he would find the fresh water spring. If he had offended the gods, the spring would disappear. By the time he reached Kahakuloa, all would be peaceful.

Perhaps this journey through the uplands to Kahakuloa was an ancient test of courage, but these place names still remain on maps today.

For those who took the coastal route to Kahakuloa, the journey began at Pu'u Kaeo, where offerings and prayers were made for knowledge to make the trip safely. By the time they reached Lae Kanounou the winds of Kalaloaloa were fierce and the traveler had to pass Nakalele Point without being detected by the supernatural sea dragon that could pull him into his cave.[271]

The Nakalele Blowhole is still the cave of Nakalele who flashes his rainbow colored scales as a warning to the unsuspecting traveler not to come too close. Even today, there are reports of people standing too close and being sucked into the blowhole and drowning.[272]

HONONĀNĀ

Bay and Valley of Protection and Watchful Care

Past Nakalele Point is Hononānā Gulch and Bay, last of the six *Hono*. Near the shore and *mauka* (inland) of the streambed, sits the unusual pentagonal-shaped Heiau Ho-na-nānā (*Ho* – to give; *na* – calm; *nānā* – to care for, observe). A large five-sided *heiau*, its walls are six feet thick and eight feet tall.[273] Hononānā is the land and bay of protection and "watchful care."[274]

PŌHAKU KANI

Past Heiau Honanana, just before the 16-mile marker, is the large stone Pōhaku Kani (Sounding Stone or Bell Stone) standing alone in pastureland. Some say the *menehune* (legendary little people) rolled this stone here so that its messages could be heard in the valleys and temples from Kahakuloa to Honokowai to Moloka'i. When tapped with the palm of the hand, the bell-like tones were carried on the winds and messages were sent quickly across the steep terrain, warning of approaching canoes. The *mana* of the messenger was essential in connecting with the spirit of the *pōhaku*.[275] People have beat the boulder with stones, trying to make it ring, but you must be a *kama'āina* (native born) with *mana* (spiritual power) to make Pōhaku Kani ring like a bell.

There are several Pōhaku Kani on other islands also. Around the large boulder are scattered several smaller stones, once called Na Kua o Hina, the tapa-beating anvils of Hina, mother of Māui, who raised her four sons at Ka-haku-loa, The Eternal Master Valley[276].

KAHAKULOA VALLEY

Ka-haku-loa (the Enternal Master) Valley

Kahakuloa Valley was made famous in antiquity as having received the sacred waters of the gods Kāne and Kaneloa.[277] On the upper trail between Kahakuloa and Honokōhau is Wai a Ke Akua, the Water of the God, a sacred spring marking this as an Ala o na 'Akua, or Trail of the Gods. These were the first trails followed by the Ka Po'e Kahiko, the people of old. Today Wai a Ke Akua is still on the maps.[278]

Kahakuloa Valley (The Eternal Master Valley) rises 4,480 feet to 'E'eke Crater, hiding in the mists. Gushing from 'E'eke are the headwaters of Kahakuloa and Honokōhau streams, that once fed thousands of *lo'i kalo* terraced on the valley floors. An early home of the Maui *Kiha* spirits, 'E'eke has always been protected by the *mo'o* against those who would desecrate the waters of streams, springs, taro patches, fishponds, and gardens. Comparable to the sanctity of Hālawa Valley on Moloka'i, Kahakuloa Valley has always been sacred. Once a very large village of fishing, farming, and sacred temple rites to Kāne and the Kiha, it remains "the Eternal Master Valley" of Hawaiian descendants.[279]

On the west side of the valley, just *mauka* (inland) of the old school, is Heiau Kāne'a'ola. Its north and west walls terrace upward 20 feet and its south wall is 6-feet thick. Its purpose is lost to time but its name is not. (*Kāne* – the Creator; *'a* – in the manner of; *ola* – life, health). A temple to Kāne, the life-affirming deity from whom Maui's highest ranking *ali'i nui* claimed divine descent. Half a mile up the valley on the east side of the stream is Heiau Kuewa. Its walls and platforms have been altered long ago for use as property lines. Near the site is a trail leading up the cliff to the remains of Heiau Pakao (*Pa* – having the quality of, *kao* – to throw a spear). Sitting at the edge of what was once a pineapple field, this *heiau* looks over Kahakuloa Valley from the top of Kuakini Ridge.[280]

A very large Pōhaku o Kāne still stands in the back yard of a home built on the foundation of what was once Heiau Keahialoa, just east of the stream.[281] Jutting up from the ground it measures seven feet tall, six feet wide, and narrows at the top. All villages once had sacred stones called Pōhaku o Kāne, personal altars where one or more families regularly prayed and made offerings. This one is exceptional in its size, which is probably why it is still in place after hundreds of years, a true relic of the ancient religion standing undisturbed in "the Eternal Master Valley."[282]

PUʻU KOAʻE

Kahekili's Leap

At the headland of Kuakini Ridge (*Kua* – tapa-beating anvils; *kini* – multitude, many) looms the massive 636 foot-high cinder cone Puʻu Koaʻe. Named for the *koaʻe* birds who inhabit its tall cliffs, it once sheltered the Koʻa of Kāne-halaʻoʻa along its shore where the sea deities were worshipped.[283] King Kahekili, The Thunderer, often dove into the sea from Puʻu Koaʻe in the sport of *lele-kawa* (cliff diving), and made this landmark famous as "Kahekili's Leap."

ST. FRANCIS XAVIER CATHOLIC MISSION

St. Francis Xavier Church

Standing sentinel on each side of Kahakuloa Village are the early missionary churches of this devout community.

HAWAIIAN CONGREGATIONAL CHURCH

Kahakuloa Church

MĀUI, THE SUN SNARER

Olowalu Petroglyph

In the annals of Hawaiian lore, Kahakuloa was also famous as the home of the demi-god Māui-a-ka-lana, The Sun-Snarer.

Among the pantheon of Polynesian demi-gods, Māui is probably the best known. Legendary in chants and stories throughout all the islands of the Pacific, Māui was known as a trickster with supernatural powers he used to help mankind. He fished up the Hawaiian Islands, raised the sky, gave people the secret of fire, and snared the sun to slow it down. Tales of his antics can be found throughout the Hawaiian Islands with the same story localized for each island. Dr. Katharine Luomala's *Maui-of-a-Thousand Tricks* (Bishop Museum Bulletin, 1949) is a definitive 15-year study of this Polynesian demi-god whose fame stretched across the great Pacific. Dr. Luomala found that tales of Māui were told among the commoners, as opposed to the creation stories which were restricted to the priests and the ruling class. She noted how the Māui myth had localized variations in different island groups to fit the needs of various societies.

On Maui, stories of the demi-god's deeds are localized in West Maui and East Maui. The West Maui version told here is similar to Fornander's account in his *Collection of Hawaiian Folklore,* but is even more "localized" through the oral traditions of the many *kūpuna* (elders) with deep ties to the land.

Māui is found in the genealogical descent of almost every Pacific Island royal family, including Hawaiian royal families.[284] Māui appears in the Ulu genealogy of the *Kumulipo* as a direct ancestor from Wākea. In the Fifteenth Era of the *Kumulipo*, Māui and his three brothers are born of Hina-a-ke-ahi and Akalana and his many heroic deeds are listed.[285]

He becomes known as Māui-a-ka-La-na (in-the-manner-of-the-sun, *La*) after he snares the sun. Māui Akalana appears in the 22nd generation of Wākea and Papa in the Ulu genealogy.[286] From here the *mo'okū'auhau* (genealogy) goes down to Hema in the 28th generation and on to Haho of the 40th generation. From Haho descend the great chiefs of Maui down to Kahekili.[287]

MĀUI AKALANA

Māui Akalana was the youngest of four sons born to Hina and Akalana above Kahakuloa. His older brothers were twins named Māui Mua and Māui Hope, and Māui Ki'i Ki'i. Living in the wet uplands of Kahakaloa, Hina often wept because the sun moved so fast across the sky that her *tapa* would not dry.

Māui determined he would slow the sun so the days would be longer for *tapa* makers, fisherman, and *kalo* (taro) planters. His friend MoeMoe, the god of Sleep who lived in a cave at Pu'u Keka'a, mocked him as a fool for taking on such a task. Māui vowed to achieve his goal and to put MoeMoe to death for mocking him. To avoid being followed, Māui leaves Kahakuloa and travels a zigzag trail around the west side of Mauna Kahālāwai and stops at Waihe'e to rest. Here he meets the kindly fisherman Kanaka-o-kai, Man-of-the-sea, and tells him of his desire to slow the sun so the days could be longer for everyone. The fisherman told Māui he would teach him how they used the *ko'olau* (hold-fast) net to encircle and snare the fish.

Hawaiian Fisherman. Woodcut print by Charles William Bartlett, 1920, Library of Congress

He showed him how to cast the net in the *hanakauhi* method and then gave Māui his *ko'olau* net and wished him a successful journey.

Māui looked toward the mountain and saw a *puka* (hole) in the cloud bank where he could see the *lani* (heavens), so he followed it to the summit of the big mountain and there he could see across the crater floor to its highest peak. Māui knew he must reach this eastern peak before the sun's first rays broke through the deep gorge cut into the crater's side. He struggled down a treacherous footpath, crossing the crater floor carrying his *ko'olau* net and climbed to the top of this highest peak. Here he sat, waiting for Lā (the Sun) to rise up. Soon the first rays broke through the valley floor and up into the crater where Māui cast his net in the *hanakauhi* manner as the fisherman had taught him. The net entangled the rays and held fast! Māui asked Lā, the Sun, to slow down so Hina could dry her *tapa*. Finally Lā agreed he would travel slowly half the year.[288] "And Winter won the Sun, so Summer was won by Māui," as told in the *Kumulipo*.[289]

To commemorate this glorious deed, Māui named the great mountain with the crater as Alehe a Kalā the Sun Snarer. The peak from which he cast his net he named Hanakauhi, and the deep gorge he named Ko'olau for the hold-fast net he used. The dangerous footpath he struggled across to reach his peak he named Hele-mau (travel steadfastly) Trail, and the path he took from Waihe'e to the mountain summit he named the Pukalani (Opening to Heaven) Trail.

Returning to the west side of the island, Māui sought out MoeMoe, the god of Sleep, and kept his promise by turning him to stone at Keka'a. Pōhaku MoeMoe still lives at Ka'anāpali.[290]

The zigzag trail from Kahakuloa to Waihe'e via Lahaina became known as Ke alanui kīke'eke'e Māui, "the zigzag pathway of Māui."[291] The Hele Mau Trail across the Ko'olau Gap to Mauna Hanakauhi is misspelled as Hale Mau'u (grass house) on maps, which Mrs. Ashdown said makes no sense.[292]

Hanakauhi is the highest peak on the sunrise side of Haleakalā at 8,910 feet, and just below it sits Mauna Hina (perhaps named for Māui's mother, Hina of the Tapa) and then Pu'u Kumu (foundation, origin, teacher). The Helemau Trail still passes at the base of these three peaks.

The Sun Snarer Mountain (Aleheakalā) becomes known over time as Haleakalā (House of the Sun) and Māui becomes known as Māui-a-ka-la-na (Māui of the Sun) and is recorded on the Ulu-Hema genealogy in the 22nd generation from Wākea and Papa as Māui Akalana, the famous demi-god who slowed the sun for mankind.[293]

This is the West Maui version of Māui the Sun-Snarer, told to Mrs. Ashdown in the early 1900s. As just one of many versions, the place names in this story *are* the *wahi pana* (storied places). Mrs. Ashdown always said, "Translate the place names correctly and the story they were left to tell, is told."

APPENDIX: ʻĀINA KAULANA

Inez MacPhee in her traveling ensemble, 1907

The source and inspiration for this book is Maui County Historian Emeritus, Inez MacPhee Ashdown. Born on a Wyoming cattle ranch in December 1899, "eighty miles from nowhere,"[294] she was transported, as if magically, to this Paradise of the Pacific, at the age of just eight. Her father, Angus MacPhee, World Champion Roper of Cheyenne, Wyoming, had been recruited by Eben Parker Low of Parker Ranch, to perform in Hawaiʻi's first Wild West Show that December of 1907.[295]

Arriving late in Honolulu aboard the *S.S. Alameda* on the day of the big event, Inez and her parents were greeted at the gangplank by a large crowd of people and the Royal Hawaiian Band playing "Cheyenne." They were swept up by "Uncle Eben" Low into a flower bedecked automobile that led a parade of flowered riders and people to Moʻiliʻili Park. People were shouting "Ma-ka-pee," which "Uncle Eben" explained was Hawaiian for MacPhee. It was a royal welcome to this new

Queen Lili'uokalani ascended the throne of the Hawaiian kingdom in January, 1892. She was forced to relinquish her throne and the sovereignty of her nation to members of the "Missionary Party" and American military forces in January, 1893. Her Majesty passed away in January, 1917. Hawai'i State Archives

land, indeed. They were driven up to the grandstand and taken to the royal box where, bowing low with a sweep of his hat, "Uncle Eben" introduced Inez and her parents to "Her Majesty, Queen Lili'u-o-ka-lani." Inez gave the Queen a deep curtsey and felt "knighted" when the Queen placed a golden *'ilima lei* around her neck and kissed her on the forehead. Inez and her mother were invited to sit with the Queen as honored quests while her father prepared to perform.

At the Grand Ball that evening, little Inez danced a waltz with Prince Kūhiō, Hawai'i's Delegate to Congress. Included in the many festivities of Honolulu society, Inez was even given an eighth birthday party at the Queen's home where "Aunt Lili'u" claimed her as "*hānai aloha*" (child "adopted in love") and presented her with a Hawaiian doll. For a little girl raised on the stories of Grimm's Fairy Tales, this was like a fairy tale come true, with her own real Queen.[296]

It was during this time in Honolulu that Inez fell in love with Hawaiian lore as her "Uncle Eben," a great-great-grandson of Kamehameha I, and Queen Lili'uokalani told her the sacred-name meanings of the valleys and mountains and streams they visited. The heroes of Grimm's Fairy Tales were instantly replaced with the heroes of ancient Hawai'i as Inez began to learn how to "walk in the footsteps of the ancient *ali'i*."[297]

Eben Parker Low, Grand Marshal, Kamehameha Day Parade, 1946

Uncle Eben Parker Low, raised on Parker Ranch, Hawai'i's first and largest ranch, was a great-great grandson of Kamehameha I and his wife Kāne Kapo Lei. Their daughter Kipikane married John Palmer Parker, a New England seafarer who arrived in Kona, Hawai'i in 1809 and was befriended by the King's court. Finally, in 1814 he was given permission by the King to begin capturing the wild cattle of Waimea. British Captain George Vancouver had gifted Kamehameha with a pair of Andalusian cattle from Monterey around 1793. Now, after a 20-year *kapu* (taboo), the cattle had become a wild herd causing destruction. John Palmer Parker started alone, walking into the forests of Waimea and shooting a beast, skinning it and carrying the meat and hide home. Soon he had Hawaiians in *malo* (loincloths) passing the beef sides down the mountain to be salted for sale to the passing ships. This began the cattle industry in Hawai'i.

John Palmer Parker and Kipikane built a home and raised a family on Waimea lands that grew to become the largest ranch in Hawai'i, with over 300,000 acres. Their daughter Mary Ann would become Eben's grandmother. Born in 1864, Eben was raised Hawaiian-style by his Hawaiian grandmother Mary Ann and her second husband the High Chief Waipā, a skilled canoe builder of Kamehameha's *peleleu* war fleet.[298] Uncle Eben spoke fluent Hawaiian and knew the genealogies and stories of his *ali'i* ancestors. As a rancher and a statesman he was well known throughout the state as "Rawhide Ben." Uncle Eben was a lifelong influence on Inez and she helped him write his memoirs in the last years of his life, because "you, a paniolo, will know because you lived part of it." Eben Parker Low died in January 1954 at 90 years of age.

Angus MacPhee, 1910. Bishop Museum, Baker Collection

Angus MacPhee was trained as a first rate cattleman in his Scottish father's Big Sybille country of Wyoming and was riding herd with the big cattle drives between Canada and Mexico by the time he was 12. At the age of 18 he left for a six-year stint traveling the world as a performer with the Buffalo Bill Wild West Show. He returned to Cheyenne in 1898 and married 16-year old Della Talbot, a convent-schooled young lady from a prestigious Cheyenne family, whose grandfather had literally built Cheyenne from his own brickyard, brick by brick. Soon after marrying, at the age of 24, Angus served with Teddy Roosevelt in his campaigns in Cuba, Puerto Rico, and the Philippines during the Spanish-American War. He also travelled to China and served in the short-lived Boxer Rebellion. In 1902 he went to Alaska as part of an expedition to bring relief to the snowed-in miners at Dawson. In 1903 he returned to Cheyenne and began management of several ranches.[299] In August 1907, his friend, President Teddy Roosevelt, came to Cheyenne as a guest of the Frontier Day Celebration, as well as a group from Hawai'i led by Eben Parker Low of Parker Ranch. Angus MacPhee broke the world's record for roping and became the new Champion Roper of the World. For the Grand Finale, he brought Inez down to ride in the arena, with "Uncle Teddy," "Uncle Eben," and her "Papa" leading the cavalcade.[300]

Soon Inez found herself on a ship on the stormy Pacific, and then sitting beside a real-life Queen. Her father decided they should stay in this Paradise and accepted the management of ʻUlupalakua Ranch on Maui.

Inez and her mother traveled to Maui aboard the *S.S. Mauna Kea* with the Queen and Her Majesty's party, who were disembarking at Lahaina while the Ashdowns went on to Mākena Landing. As they entered the ʻAuʻau channel approaching Lahaina, one elderly aristocratic lady tossed her head and said, "Lahaina Roads, indeed! ʻAuʻau means our entire background since the beginning of creation. How will our children remember the beauty of our island culture when even our melodious Hawaiʻi names are forgotten?"

Queen Liliʻuokalani replied, "Some will remember."[301]

As the Queen disembarked, little Inez watched as the crowd ashore fell to their knees and began wailing a greeting to their Queen in the ancient fashion. Some reached out and kissed the hem of her dress as she slowly walked to her carriage.[302]

By January of 1908, Inez found herself in a new wonderland as an eight-year-old riding the high green pastures of ʻUlupalakua Ranch with her own Hawaiian guardian, white-haired Kahu Kīnaʻu, a revered *pa-ni-olo* (cowboy) of *aliʻi* descent who appointed himself her *kahu* (guardian/teacher). Pointing down to the churning sea channels among the small islands below, Kahu Kīnaʻu began his lessons

ʻUlupalakua Ranch house

by telling her the name of the channel between Kaho'olawe and Lāna'i; Ke-ala-i-kahiki, meaning "the pathway to and from the horizon" and how it was given this name by Hawai'i-loa-ke-kowa, The Discoverer, who found these islands from beyond *kahiki-kū* (the broad horizon).

As she learned the place names, geography, history, and lore of her new home, each story would be discussed and written as part of her daily lessons. Eager to be like them, Inez learned to be a *paniolo* from her father and *Kahu* and the many *paniolo* who called her Ke Aka, "The Shadow."[303]

Her father had become great friends with Louis von Tempski, the manager of Haleakalā Ranch. Sixteen years older than her father, von Tempski had come to Hawai'i at the age of 19 from New Zealand, where his father Major Gustavus von Tempski had died a hero in the Māori Wars.[304] Louis married Amy Wodehouse, daughter of the British Commissioner to Hawai'i Major James Hay Wodehouse. At her untimely death in 1909, Louis was left with four children to raise: Armine, Gwen, Lorna, and Errol. Sixteen year-old Armine was assisted with the younger children by her older cousin 'Āina Wodehouse, who was also a cousin of Crown Princess Ka'iulani, heir to the throne of Queen Lili'uokalani.[305]

Louis von Tempski and his daughter Armine, author of Born in Paradise

Inez MacPhee and Lorna von Tempski at Kapalaea, c. 1916

At the age of 15 Inez came to live at Kapalaea, the Haleakalā Ranch manager's house in Makawao, with "Uncle Von," who renamed her "Jackie" (English for Keaka), and his *paniolo* children. Jackie, Lorna, and Errol would ride horseback to Maui High School together and became, literally, blood brothers. Fluent in Hawaiian, Uncle Von knew the lore of Haleakalā and its crater and took his *paniolo* children on cattle drives and hunting trips into the crater and to all the neighboring ranches.[306]

At a time when Hawaiian history was not taught in the schools, the von Tempski children were all schooled in W. D. Alexander's *A Brief History of the Hawaiian People*, 1899, as well as Charles Baldwin's 1908 *Geography of Hawai'i*, as required reading in their home studies.

Inez's early life was spent as a *paniolo* (cowboy), riding and roping with the cowboys of Maui's great ranches: 'Ulupalakua, Kaono'ulu, Haleakalā, Kaupō, Kahikinui, Kahakuloa, and Kaho'olawe. By the time she was twenty, she had ridden most all the trails of these ranchlands, including Haleakalā Crater and its many trails, and knew many of their landmarks and legends.[307] In 1919, she and the von Tempskis discovered unknown stone structures in the bowl of a 300-foot tall cinder cone on the bottom of the crater. Inez took photos and Bishop Museum sent archeologists to inspect the structures.[308] This led to the 1922 Survey of Haleakalā Crater by Dr. Kenneth Emory, revealing over 58 ceremonial platforms and terraces throughout the "House of the Sun."[309]

Inez helping with the branding at Kahikinui

Frank and Harriet Baldwin, Inez, and her father Angus, Kahikinui House

Ku He'eia Bay, Kaho'olawe, 1919

Paniolo Moki Medeiros of Kula recalled those early days in a 1982 Maui News article about Inez when she was Grand Marshall of the Makawao Rodeo Parade; "That was one of the few *paniolo wahines*. I've seen her ride—beautiful." She also trained and exercised race horses. In 1920 she won the Silver Cup in the Cowgirl Race at Kahului Fairgrounds riding Lahotan Water.[310]

Through her father and Uncle Von, Inez became acquainted with the Baldwin Family, owners of Haleakala Ranch and Hawai'i's largest sugar plantation, H. C. & S. Company. In 1922 Harry Baldwin, son of H. P. Baldwin, became partners with her father in the Kaho'olawe Ranch Company, and soon he became "Uncle Harry" to Inez.[311]

In 1918 Angus MacPhee obtained a 21-year lease from the Territorial Government on Kaho'olawe, an island barren of plants and trees, devastated by thousands of goats running wild on land whose soil was worn down to hardpan. Having closely observed Kaho'olawe and the rain patterns of the clouds from 'Ulupalakua, MacPhee believed he could bring the island back to life and make it into a fine cattle ranch. People thought he was crazy. When the Kona winds blew, large red dust clouds would rise from Kaho'olawe.

A condition of the lease was that MacPhee had to rid the island of goats and sheep within four years. Inez considered herself a partner with her father in this monumental endeavor and with the help of the von Tempskis and *paniolo* like Jack Aina and Manuel Pedro, they shipped and sold over 13,000 head of sheep and goats to Maui and Honolulu. Many others that could not be corralled had to be shot. They won the lease for another 30 years, but MacPhee had gone bust financially, so in 1922 his friend Harry Baldwin came in as his partner.[312]

'Ulupalakua School children, 1923

At the age of almost 20, Inez was sent to finishing school at Dana Hall in Massachusetts. Returning in 1922, she began teaching at 'Ulupalakua School and would occasionally take time off to work to be with the *paniolo* on Kaho'olawe or to participate in the brandings at nearby ranches. In 1926, Harriet Baldwin asked Inez to go to U.C. Berkeley and take the courses necessary to become Executive Director of the Girl Scouts. Here she met an English exchange student, Charles Ashdown, whom she married and brought back to Maui with her in 1928. "Uncle Harry" Baldwin gave Charles a job as field timekeeper at Honolua Ranch. Inez taught him how to ride a horse. Under the management of David T. Fleming, Honolua Ranch became an idyllic home for Inez to raise her two young sons.[313]

Now a mother and no longer a working *paniolo*, Inez became involved in many civic organizations. Around 1933, Mrs. Fleming asked Inez to take over as president of the Women's Club. Her first task was to get the dump at the end of town covered over. The county was using Wai-o-kama fishpond as a dump and it was unsightly. Mrs. Ashdown went to the home next to the dump and found Bill Ka'ae, her old friend from Wailuku days when he was a politician. He introduced his wife, Alice Ka'ehu-kai Shaw Ka'ae, "we all call her Aunty Kai, short for Ka'ehu-kai, the name given her by Kamehameha V." Aunty Kai knew why she was there and told Inez that it was the county's responsibility to cover over their rubbish. Inez told her she would talk to her friend Harold Rice (of Kaonoulu Ranch) who was on the Board of Supervisors. "I've roped cattle with him, I'll ask him to take care of it." Aunty Kai said "Fine," and took her by the arm as they walked Mākila Beach and she told Inez the story of the King's old fishpond. Aunty Kai had been a Lady-in-Waiting to Queen Liliʻuokalani and their shared love of "Aunt Liliʻu" bonded them immediately. Mrs. Ashdown spoke with Mr. Rice and with the help of Jack Moir at Pioneer Mill and David T. Fleming of Honolua Ranch the county brought soil from their plantations and filled in the dump.[314]

Always involved in many civic organizations, she began writing a column for the *Maui News* and the *Star Bulletin* in 1936. Beginning with *A Brief History of the Hula,* she wrote about the Hawaiian culture and history she had learned from her Hawaiian *kūpuna* (elders). When she wrote and published a chant extolling the famed hula dancer, ʻIolani Luahine, the Maui Hawaiian Women's Club gave her the name ʻĀina Kaulana, engraved on a gold bracelet, meaning she brought pride to the land.[315]

It should be remembered that at this time the Hawaiian language, culture, and history had been suppressed for decades. The Hawaiian language had been forbidden in the schools since 1850, and since the U.S. Annexation in 1898 the emphasis was on becoming "American" and to learn to "swim with the overwhelming tide." Inez taught a class of 36 students at ʻUlupalakua School, 18 Hawaiian and 18 Japanese children of ranch employees. When she asked them in 1925 what race they were, they all replied "American."[316]

While she continued writing she also began speaking at Hawaiian Civic Clubs, urging them to preserve their historic sites, look up their genealogy, and to teach their language before it was too late. Soon the Maui Hawaiian Women's Club and the West Maui Hawaiian Civic Club invited Mrs. Ashdown to become an honorary member. Under her direction, they formed historic sites committees and began identifying the *wahi pana* (storied places) of their districts.

Prince Kūhiō had started the Hawaiian Civic Clubs before his death in 1922, and many of the elders in the West Maui Hawaiian Civic Club were charter members. Among them were Hawaiian matriarchs of *aliʻi* families and Ladies-in-Waiting to the Queen, who remembered the time of the *aliʻi* and their illustrious past and wanted it preserved before it was too late. After decades of suppression of their language and culture, at a time when Hawaiian history still wasn't taught in the schools, these *kūpuna* of the Hawaiian Civic Clubs began their own Renaissance of preserving their Hawaiian heritage.[317]

Among the matriarchs of the West Maui Hawaiian Civic Club were the Shaw sisters, Alice Kaʻehu-kai Shaw Kaʻae and Mary Kawaiʻele Shaw Hoapili. Born on Molokaʻi in the mid-1860s, they were adopted by the bachelor King Lot Kamahameha V who named them Kaʻehu-kai (The Sea Mist), and Kaʻwaiʻele (The Deep Waters), at his home "Māmala" at Kau-na-ka-ha-kai, as Kaunakakai was then known.

Alice Ka'ehukai Shaw Ka'ae, "Aunty Kai"

From *ali'i* heritage, their mother Lahela was born in Lua'ehu in 1832 when her family owned all of Ka'anāpali and much of the Lua'ehu district of Lahaina. Lahela's father was Nālehu, a grandson of Hawai'i King Alapa'inui, and *konohiki* of Waiokama lands.[318] Lahela's uncle Nahaku was historian to King David Kalākaua.[319] Their father was Kanaina, a relative of King Lunalilo. Later their mother married Patrick Shaw, who served as Governor of Moloka'i for King Kamehameha V. His father, William Shaw, had also served His Majesty as governor of Moloka'i and was also married to an *ali'i* Chiefess. Shaw Street in the Lua'ehu area of Lahaina was named for William Shaw.[320]

Both Aunty Kai, short for Ka'ehu-kai, and Aunty Ka'wai'ele lived on their family land at Mākila beach, next to the Wai-o-Kama fishpond. These noble women were raised among the royal court and were witness to the last days of the monarchy. The sisters were close family to King David Kalākaua and his sisters, Princess Lili'uokalani and Princess Likelike. Both served as Ladies-in-Waiting to Queen Lili'uokalani.

Aunty Kai and Aunty Kawai'ele would lead Inez (and sometimes other civic club members) on expeditions up and down the coastline. Aunty Kai would say "Halt" and Inez would stop and write down everything she said about the place.[321]

Another matriarch of the civic club was Annie Hinau, also of *ali'i* heritage, her mother Ka-'ohu-lani was a Lady-in-Waiting to the Queen. Annie was born in 1900 and was adopted by her maternal uncle John Kekinomakani. His mother was Wahine-nui Napua of ʻUlupalakua. Annie married Junior Hinau, who was born and raised in his grandfather Napaepae's house by the King's Hale Piʻula in Luaʻehu.[322]

Another active club member was Alice Makekau Aki. Born in 1893, Alice attended Hale Aloha and the Luaʻehu school as a child. Her mother, Mima Makekau, was of the priestly class of *kāhuna lapaʻau* (natural healers). Her father was Alfred Keʻliʻi Makekau o Nuʻuanu.[323] Alice remembered seeing the Kiha Wahine Mokuhinia when she was a child. She said the *moʻo* looked like a mermaid with long black hair.[324]

Others active in the work of the historic sites committee of the West Maui Hawaiian Civic Club during this period included Annie Ako, Lahela Reimann, Hattie Kanaka-o-kai Yoshikawa and her sister Elizabeth Namauʻu, Mary Chan Wa, William Saffrey, Robert Saffrey and Winnie Saffrey Sanborn, Anaka-luahine, John Lihau and his brother George Kaaehui, Tommy Kekona, John Kapa-ku, Reverend John Kukahiku, Pilahi Paki, Kaniho Humehume, Sam Makekau, James Kahahane, and Manuel Silva, all *kamaʻāina* contributing to the knowledge of their specific region of West Maui.[325] They wanted the *wahi pana* (storied places) preserved and the oral *moʻolelo* (stories, history) of West Maui recorded into written history, which they trusted Mrs. Ashdown to do.

They also became activists. Long before there was any official system of historic preservation, through their direct action, the West Maui Hawaiian Civic Club saved many of the historic sites in West Maui from total destruction.

When a new labor union wanted to tear down the old Baldwin House to build a bowling alley, the civic club members sent Mrs. Ashdown to speak to Frank Baldwin and ask him to save it. She told him "You can't allow these *malihini* (foreigner) union people to do this. The *kamaʻāina* (native born) love the memory of the Baldwins and appreciate all they've done for Maui. You can't let this house go. If you fix it up, they'll have no excuse to tear it down." Frank Baldwin spent $19,000 of company funds restoring the building and turned it over for use by the Girl Scouts.[326]

When Hale Paʻahao (the old prison) was in danger of falling down, again with the support of the West Maui Hawaiian Civic Club, Mrs. Ashdown went to her old friend Harold Rice, still County Supervisor, and asked for the county's help. Jack Moir of Pioneer Mill and David T. Fleming of Honolua Ranch pitched in and Hale Paʻahao was fixed up for use as Boy Scout Headquarters.[327]

There would have been no Hale Paʻi (Printing House) at Lahainaluna to preserve except that Sam Moʻokini, then President of the West Maui Hawaiian Civic Club, and Mrs. Ashdown approached Frank Kennison, Principal of Lahainaluna, and asked him to help them save it from destruction. Mr. Kennison was able to obtain funds for the first time in 25 years and they built new classrooms, shored up the walls

Jack Aina shipping cattle, Ku He'eia Bay, Kaho'olawe

and used a portion of Hale Pa'i as a museum for many years, until Mr. Kennison left the school. Today Hale Pa'i is the only original building left at Lahainaluna.[328]

In 1937 and 1938 Inez wrote full-page articles in the *Honolulu Star Bulletin* and the *Maui News* about the history and the progress of Kaho'olawe Ranch Company, by now a productive cattle ranch. They had mixed Australian salt bush and Red

Kaho'olawe visitors, 1938

Top seeds with the *pili* grass that had started to come back and produced a ground cover that stopped the erosion of the island's red soil and provided feed for over 700 head of black Poll Angus cattle and a herd of thoroughbred mares and foals. They had planted over 5,000 indigenous seedling trees and windbreaks of eucalyptus. In less than 20 years, Angus MacPhee had made a working ranch of the once desolate Kaho'olawe.[329]

In 1939, with war looming, Senator Baldwin and Angus MacPhee, in a spirit of patriotism, leased the southern tip of Kaho'olawe to the U.S. Army for bombing practice. Almost 1,000 head of cattle were removed to Baldwin's Grove Ranch on Maui. After the Japanese attack on Pearl Harbor on December 7, 1941, the entire island was appropriated and used for target practice throughout the war, and beyond.[330] Due to the numerous unexploded ordinances, the island was never returned to the Kaho'olawe Ranch Company.

At the outbreak of WWII, Inez went to work as the only civilian communications employee at the USS Naval Air Station Pu'unēnē in 1942. Because of her vast knowledge of island geography, Mrs. Ashdown was hired as switchboard operator and served through 1946, providing a consistent flow of information among all the military units on Maui. Housed in concrete Building 74 on the flatlands between Wailuku and Kihei, it was a grueling job that ended with a stint in the hospital. At the end of the war the U.S. Navy honored her with their Meritorious Award for her years of outstanding service. In 1946, the commanding officer at NAS Pu'unēnē took Inez and some officers of the U.S. Army Corps of Engineers to Kaho'olawe. Mrs. Ashdown was so shocked by the island's total destruction that she could not go ashore.[331] In 1979, Mrs. Ashdown wrote her book, *Recollections of Kaho'olawe.*

Mrs. Ashdown went back to historic sites work with the Hawaiian Civic Clubs and began writing again for the *Maui News* and the *Honolulu Star Bulletin.* The *kūpuna* of the West Maui Hawaiian Civic Club had accomplished a lot, but now they wanted a more comprehensive approach to historic preservation and they wanted the Territorial Government to participate. In 1950 they invited a group of

Mrs. Ashdown at NAS Pu'unēne

Territorial Government legislators to the old Pioneer Theater in Lahaina to listen to their concerns. Sam Moʻokini, a language and music teacher at Lahainaluna, spoke in Hawaiian and Mrs. Ashdown spoke in English. Their plea was heard and in 1951 the Territorial Historic Sites Commission was created under Governor Long. Each county was appointed a Historic Sites Commissioner to organize the identification of historic sites.[332]

Mrs. Ashdown was appointed the Historic Sites Commissioner for Maui. She organized historic sites committees all over the county with Hawaiian Civic Club members and people knowledgeable in each district. With a list of heiau sites compiled from archeological surveys taken by Thomas G. Thrum, J. F. G. Stokes, and Winslow Walker between 1909 and 1928, and other sites known to committee members, they began to search them out and inventory them with modern tax map key numbers for permanent identification and hopefully, preservation. Many important sites and their histories were saved this way, including the sites of Olowalu: Puʻu Kīleʻa and the Olowalu petroglyphs, the Heiau Kaʻiwaloa, and the Heiau Hekiʻi flanking ʻUkumehame Valley. Found still resting on the lands of Pioneer Mill Company, Jack Moir, then manager of the company and a member of the Historic Sites Committee, agreed to set aside these sites for preservation, decades before the state had any historic preservation program.[333]

During this same time Mrs. Ashdown was also working with Hawaiians in Central Maui, Hāna, Kaupō, and South Maui to map the sites in their districts.

In 1956, Mrs. Ashdown was a founding member of the newly organized Maui Historical Society, now also involved in the preservation efforts. In 1957 she was instrumental in obtaining the Old Bailey House from Wailuku Sugar Company for use as a museum and headquarters for the Maui Historical Society.[334]

Jack Moir with Bryant Cooper at Heiau Kaʻiwaloa, 1956

Sites work at Heiau One'uli in Mākena

As tourism became the focus of Maui's economy in the early 1960s, the importance of preserving the historical character of Lahaina received a renewed interest from the Government, and in 1961 Lahaina Historic District 1 was created, giving the historic sites saved within Lahaina Town permanent preservation status. In 1962 the state legislature appropriated funds for the restoration of Lahaina and the Lahaina Restoration Foundation was created to manage the restoration and sites within the Lahaina Historic District. In 1962 and 1963, the first two hotels were built at Ka'ānāpali: the Royal Lahaina Hotel and the Sheraton Maui Hotel.

In 1968 Mayor Elmer Cravalho hired Mrs. Ashdown as County Historian to head the new Hui Hana Malama, meaning Group Working to Protect, named by the Reverend Edward Kapo'o. Funded by the Mayor's office, this group continued locating and marking historic sites island-wide.[335] It was Hui Hana Malama that rediscovered and uncovered the village of Wailea from decades of overgrowth. Mrs. Ashdown hired Dr. Kenneth Emory of Bishop Museum to survey the site and in

Hattie Kanaka o kai Yoshikawa and Sam Po at Ma'onakala Village above Ahihi Bay. Hattie, Sam, and Inez did field work together at Kahikinui, Kaupo, Hana, Waiehu and Waihe'e.

Early site work at Heiau Pi'ilani Hale in Hana

July of 1970, Dr. Emory publicly declared the village so significant that it should be preserved as a "living museum."[336]

As late as 1981 Mrs. Ashdown was still lobbying A&B and Wailea Development Company to give the site preservation status. Finally, on December 2, 1981, Richard Cox of Wailea Development Company wrote to assure her that the site would be preserved and entered as such on the Kihei Community Plan.[337] With their support, on prime Wailea real estate, ancient Wailea Village is preserved today in the 21-acre Palau'ea Cultural Preserve.

In 1970 Mrs. Ashdown wrote her book *Ke Alaloa o Maui* (The Broad Highway of Maui) describing the historical sites work done to date of the Hui group. She continued her field work until 1975, when she began researching and writing "The Districts of Maui" for the county. Mayor Cravalho wanted the names of beaches, streams, and valleys and the history of every district on Maui. Charles Keau and Betty DuBois finished the field work of the Hui and finalized an inventory of the sites of Maui by 1976. By 1979 there were over 100 sites on Maui and over 500 sites statewide listed on the State Historic Register. Sadly, on March 21, 1980, all of the sites on the State Historic Register were removed from the list because the Hawai'i Historic Review Board had not properly notified property owners prior to the listing. A very small percentage of sites on private property have ever been re-registered.[338]

During the 1970s, all the major resort companies hired Mrs. Ashdown to name their beaches and identify their historical sites, as well as providing the ancient history of their resort lands. She did this for Alexander & Baldwin at Wailea, for Maui Land & Pine at Kapalua, for Seibu Corporation at Mākena, and for Amfac at Ka'anāpali.[339]

Due to ill health, Mrs. Ashdown retired from the county in 1978, but she did not stop advocating for the preservation of Maui's cultural heritage.

Hale Pa'i

Still fighting for Hale Pa'i's survival in October of 1979, she led an aggressive lobbying effort to force Governor Ariyoshi to release funds appropriated for the restoration of Hale Pa'i, a registered National Historic Site since 1976. Operating as a museum until it was condemned as unsafe by the state in 1975, it had since been abandoned and left to crumble on state land.

As the funds needed for repair were about to "lapse" for the second time, Inez and her friend Jan Dapitan of the County Department of Human Concerns sounded the alarm. Mrs. Ashdown enlisted the help of her friends: Lt. Governor Jean King, Mayor Hannibal Tavares, the Historic Hawai'i Foundation, and the press. The state had told them that Hale Pa'i was not high enough on the priority list to get the necessary $300,000 needed for restoration.[340] They strongly disagreed. In November 1979, Mrs. Ashdown's friends at *The Maui Sun* published a full-page article entitled "Historic Building is in Danger of Collapse," exposing the controversy to the public. Soon a coalition known as Friends of Hale Pa'i was formed as a fund raising group. The Hawai'i Newspaper Agency pledged $25,000 to the cause, bringing statewide attention to the plight of Hale Pa'i. In February 1980 Mayor Hannibal Tavares wrote to Governor Ariyoshi urging him to make the funding available, then issued a news release of the letter, adding; "It is a sad commentary to modern civilization that we sit and watch a truly historic site such as Hale Pa'i crumble to the ground. Hale Pa'i must be preserved."[341]

In March 1980 *The Maui Sun* published an update on the efforts to get the funding released, which the state still refused to do. Jan Dapitan mentioned that the Western Publishers Convention was being held the next month in Honolulu and was planning to take members of the Friends of Hale Pa'i to share the plight of the first publishing house east of the Rockies. In April 1980, the *Honolulu-Star Bulletin* devoted the front page of their Sunday Today section to Hale Pa'i, and Historic Hawai'i Foundation published a 4-page special on Hale Pa'i in their monthly newsletter, announcing that the state had just committed to releasing $50,000 to stabilize the walls. But that wasn't good enough.

By now Mrs. Ashdown had called on Lahainaluna alums and they wanted Hale Pa'i restored for the 150th Anniversary of Lahainaluna, to be celebrated the following year. Finally, the pressure was too much, and the state relented. With the help of Lt. Governor Jean King, on September 10, 1980, an Inaugural Blessing for the Restoration of Hale Pa'i was held at Hale Pa'i.[342] With a $145,000 check in hand, Susumu Ono, State Historic Preservation Officer, told the community leaders at the blessing ceremony that this was the first installment of a $300,000 appropriation for the restoration of Hale Pa'i, calling the effort to bring about its restoration "unique," and that it "has been one of the fastest moving projects I've seen in a long time."[343] Inez Ashdown, representing The Friends of Hale Pa'i, said it was "one of the happiest days of my life."[344] The original building was fully restored in 1982, utilizing $290,000 in state funds and $58,000 in donated funds. It has operated as a museum since then under the management of the Lahaina Restoration Foundation.[345]

In 1983, through phone calls and letters, Mrs. Ashdown enlisted the help of the U.S. Army, the State Parks Director, and the Maui Contractor's Association to finally have the unsightly WWII U.S. Army water tank removed from the site of the Heiau Hale Ki'i and Pi'ihana at Paukūkalo in Wailuku, a National Historic Site. In 1956, these heiau had been cleared by Historic Sites Commission members and volunteers under the direction of Dr. Kenneth Emory of Bishop Museum. In 1959, with the help of Mayor Eddie Tam, Mrs. Ashdown and the Historic Sites

World War II water tank, July 1983.
Maui News photo by Jill Engledow

Commission were able to dedicate the heiau for preservation, despite the fact there was still no official process for preservation. In 1971, Mrs. Ashdown used *Hui Hana Malama* funds to have Bishop Museum repair and restore the *heiau.*

Others had tried to have the unsightly tank removed, including Mayor Tavares, but Mrs. Ashdown had a way of persuading people to do the right thing. The abandoned tank was removed by the U.S. Army Reserve in 1983.[346] Finally, in 1985 the *heiau* were formally listed on the National and State Historic Site Registers.

Whatever the focus of her preservation efforts, Mrs. Ashdown always had the sound backing of the Hawaiian Women's Club, the East and West Maui Hawaiian Civic Clubs, the Ka'ahumanu Society, members of Hale o Nā Ali'i and whatever mayor was in office at the time: Mayor Eddie Tam, Mayor Elmer Cravalho or Mayor Hannibal Tavares.

In 1980, Mrs. Ashdown's years of dedicated efforts were recognized statewide when she was given the annual Preservation Award from Historic Hawai'i Foundation. She also received national recognition in 1980 when she was honored with the Meritorious Public Service Medal from the National Association of Secretaries of State. In 1982, in recognition of her efforts and devotion to historic preservation, Mayor Hannibal Tavares honored Mrs. Ashdown with the lifetime title Maui County Historian Emeritus.[347]

In 1985 Mrs. Ashdown began appearing in a weekly program called *Hula Kahiko o Hawai'i* at The Shops at Kapalua. With Cliff Ahue's *hula hālau* dancing *kahiko* (ancient) style hula to *oli* (chants) written for great chiefs or deities like *Pele,*

Mrs. Ashdown with Halau Hula Ho'oula O Ka'ula

Cliff 'Paliku' Ahue performing in the courtyard at The Shops at Kapalua

Mrs. Ashdown would talk story about the subject of the dance, between the dances, in the traditional Hawaiian fashion of passing on the *mo'olelo.* Even on a bad day she was never at a loss for words. She quickly became *Tūtū* (Grandmother) to the *Hālau* and to all of us who worked at The Shops.

People flocked to the show, and afterward visitors would line up to have Mrs. Ashdown autograph their copy of *Ke Alaloa o Maui*. She continued this show until 1989, when her health began to slow her down.

For Mrs. Ashdown's 90th birthday, her community of friends came together to honor her at 'Īao Lodge. Mayor Hannibal Tavares flew in from Hāna to spend the afternoon by her side, feeding her papaya. Cliff Ahue's *hula hālau* came to dance and Poki's band came from Lahaina to play. Representatives of the Protect Kaho'olawe 'Ohana came to honor her and brought her soil from Kaho'olawe. The Children of the Rainbow came and sang her songs and brought her a *haku lei* (woven head *lei*) made of forest buds and ferns from 'Īao Valley to crown her head.

Honolulu entertainer and radio host Napua Stevens came with her family, and she and Aunty Emma Sharpe of Lahaina danced the *hula* with *Tūtū*. For several hours folks stepped forward one by one to pay tribute to Mrs. Ashdown with a song, with a dance, with a story, with a poem. Mayor Tavares spoke about his admiration for her since small-kid time, when he saw her as a *paniolo* in Makawao. He spoke about her dedication to the preservation of Maui's cultural heritage, and how invaluable her contribution had been. He said she had a brain like a computer. Then he serenaded her with a love song. It was a loving tribute to a beloved Maui treasure. Mrs. Ashdown passed away on October 18, 1992 at the age of 92.

Mrs. Ashdown left behind a mind-boggling 50 years of written research and writings that cover most any subject regarding Maui County and its cultural her-

Mrs. Ashdown with Mayor Tavares and the Children of the Rainbow preschool

itage, including its flora and fauna. (An avid protector of the environment, she was instrumental in saving Kanahā Pond and Kealia Pond as bird sanctuaries.) A prolific writer, Mrs. Ashdown kept copies of everything she wrote and much of the research materials she referenced. From her newspaper stories of history and legends, to the many writings she did for schools and organizations, her archive of material spans from the 1930s to the 1980s and predominately includes the work she did with the Territorial Historic Sites Committee and for the County of Maui. Over four hundred subject files of her work are now catalogued and held in the archives of the Maui Historical Society at Bailey House Museum in Wailuku, Maui.

Mrs. Ashdown dancing hula with Napua Stevens and Emma Sharpe

The author with Mrs. Ashdown at her 90th birthday party

It is from this archive, as well as from a personal collection of materials given to me by Mrs. Ashdown in the last years of her life, that this book has been written. It was her wish, like that of her *kūpuna* teachers, that the *wahi pana* (storied places) be remembered and that her work be "released from the black hole of Calcutta (filing cabinets) so the *keiki o ka 'āina* (children of the land) will know their rich heritage." With this book, based on her research from the *kūpuna* descendants of the oldest families of each district, the *wahi pana* of West Maui are remembered.

Mrs. Ashdown looking at Mauna Kahālāwai from 'Ulupalakua

Notes

*All file numbers prefaced with AR11 are from the current Maui Historical Society (MHS) Index. File numbers prefaced with A15 are from the previous MHS Index.

1 Ashdown, Inez, Nā Akua o Maui, or The Gods of Maui, p. 1, (MHS) A15 11-10.

2 Ashdown, Inez, Lahaina notes, (MHS) A15 10-8.

3 Ashdown, Inez, The Shaws, 1/22/76, pp. 1, 2, (MHS) A15 15-3; Klieger, C. P., *Moku'ula*, p. 8, Bishop Museum Press, 1998: Genealogy, (MHS) AR11, 5-13.

4 Ashdown, Inez, Memoirs of Inez Ashdown,1986, as told to Jill Engledow/Lynne Ashdown, p. 98, personal collection, see also (MHS) AR11 A2 1-1.

5 Ashdown, Inez, (MHS) A15 10-1.

6 Ashdown, Inez, (MHS) A15 10-8.

7 Ashdown, Inez, Lahaina Assignment, 5/8/75; Letter to Pilahi Paki, (MHS) A15 11-6B.

8 Aina Kaulana, Memoirs of Inez Ashdown, 1986, as told to Jill Engledow/Lynne Ashdown, p. 101, personal collection, see also (MHS) AR11 A2 1-1.

9 Aina Kaulana, Memoirs of Inez Ashdown, 1986, as told to Jill Engledow/Lynne Ashdown, pp. 99–102, personal collection, see also (MHS) AR11 A2 1-1.

10 Aina Kaulana, Memoirs of Inez Ashdown, 1986, as told to Jill Engledow/Lynne Ashdown, p. 125; personal collection, see also (MHS) AR11 A2 1-1; Ashdown, Inez, Sacred Temple of Olowalu, 10/67, (MHS) A15 6-9.

11 Ashdown, Inez, Wailuku Historic District, 7/70, (MHS) A15 17-7.

12 Ashdown, Inez, Name of new park, p. 4, personal collection.

13 Ashdown, Inez, Letter to Herb Kane, p. 8, 8/22/80, personal collection.

14 Walker, Winslow, *Archaeology of Maui*, 1931, pp. 21, 22, Bishop Museum.

15 Ashdown, Inez, The Ali'i, Descendants of the Gods, Part Two, p. 5, (MHS) AR11 1-24.

16 Ashdown, Inez, Mai-ka-po-mai, personal collection.

17 Ashdown, Inez, Mo'oku'auhau, pp. 2, 3, 1974, (MHS) AR11 2-47.

18 Ashdown, Inez, 'Īao, Sacred Valley of Worthy Ali'i, p. 1, 4/18/74, personal collection.

19 Ashdown, Inez, Mo'oku'auhau, p. 3, 1974, (MHS) AR11 2-47.

20 Ashdown, Inez, Ceremonies, (MHS) A15 15-25.

21 Ashdown, Inez, Wai Ho'omana'o (Waters of Memory) 11/4/63, (MHS) A15 10-8.

22 Ashdown, Inez, Time of the Ali'i, Waters of Memory, (MHS) A15 10-8; Mauna Kahālāwai, personal collection.

23 Ashdown, Inez, Mauna Kahālāwai, personal collection.

24 Ashdown, Inez, Wai Ho'omana'o, 11/4/63, (MHS) A15 10-8.

25 Ashdown, Inez, The Kiha, p. 3, (MHS) A15 10-1.

26 Ashdown, Inez, The History of Honolua Ranch, p. 2, personal collection, see also (MHS) AR11 A2 2-60, Pu'u Kukui, p. 1, personal collection.

27 Sterling, Elspeth, *Sites of Maui*, 1998, p. 60, Bishop Museum Press.

28 Ashdown, Inez, Mauna 'E'eke, Avoiding Mountian, personal collection.

29 Ashdown, Inez, The History of Honolua Ranch, p. 7, personal collection, see also (MHS) AR11 A2 2-60.

30 Beckwith, Martha, *Hawaiian Mythology*, 1971, p. 314, University of Hawaii Press.

31 Ashdown, Inez, Kai O Hinali'i, (MHS) A15 11-6A.

32 Ashdown, Inez MacPhee, Hawai'i Nei, pp. 1, 2, personal collection; The Lahaina District, pp. 11–12 (MHS) A15 7-7B,.

33 Ashdown, Inez, *Recollections of Kaho'olawe,* 1979, p. 50, Topgallant Publishing Co., LTD.

34 Ashdown, Inez, The Lahaina District, pp. 11, 12, (MHS) A15 7-7A.

35 Ashdown, Inez, Lahaina side, 6/10/75, pp. 4, 5, (MHS) A15 10-9.

36 Ashdown, Inez, Ancient History and Legends, Lahaina side, 1975, (MHS) A15 7-7B.

37 Ashdown, Inez, Sacred Temple of Olowalu, Maui, 1967 (MHS) A15 6-9; Ancient History and Legends, Lahaina side, 1975, p. 5 (MHS) A15 7-7B.

38 Ashdown, Inez, Sacred Temple of Olowalu, Maui, 1967 (MHS) A15 6-9.

39 Ashdown, Inez, Sacred Temple of Olowalu, 1967, (MHS) A15 6-9.

40 Ashdown, Inez, Hawai'i Nei, pp. 1–2, personal collection; The Lahaina District, pp. 11, 12, (MHS) A15 7-7B; Ashdown, Inez, notes on Ukumehame, personal collection.

41 Fornander, Abraham, An Account of the Polynesian Race, 3 Vols. 1878–1885, I:23–24, 132–159.

42 Ashdown, Inez, Ancient History and Legends, Lahaina side, p. 6, 1975 (MHS) A15 10-9.

43 Ashdown, Inez, The Lahaina District, p. 11, (MHS) A15 7-7B.

44 Smith, Katherine Kama'ema'e, *Pu'uhonua: The Legacy of Olowalu,* pp. 7, 11.

45 Ashdown, Inez, History, Lahaina Area, 2/25/75, (MHS) A15 10-8, p. 2.

46 Ashdown, Inez, The Lahaina District, p. 12, (MHS) A15 7-7B; Lahaina side, 1975, p. 5, (MHS) A15 10-9.

47 Ashdown, Inez, Lahaina side, 1975, p. 7, (MHS) A15 10-9; Lahaina area, 2/25/75, p. 1, (MHS) A15 10-8; The Lahaina District, pp. 14, 15 (MHS) A15 7-7B.

48 Ashdown, Inez, Beaches, personal collection.

49 Ashdown, Inez, Fourth Lesson Mauian, 10/19/57, AR11 2-13.

50 Ashdown, Inez, The Lahaina District, p. 15, (MHS) A15 7-7B.

51 Ashdown, Inez, Lahaina Area, 2/25/75, p. 1, (MHS) A15 11-6A.

52 Ashdown, Inez, The Peacemaker, May 22, 1964, pp. 2–3, (MHS) A15 11-6C.

53 Ashdown, Inez, The Lahaina District, p. 15. (MHS) A15 7-7B.

54 Kumu Pono Associates, *A Collection of Traditions and Historical Accounts of Kaua'ula and other Lands of Lahaina, 2007, Summary of Findings,* pp. 4, 6.

55 Ashdown, Inez, The Winds, Rains and Surfs of the Lahaina District (MHS) A15 7-7D; Notecard on winds, personal collection.

56 Fornander, Abraham, *Hawaiian Antiquities.*

57 Ashdown, Inez, Lahaina Area, 2/25/75 (MHS) A15 10-8.

58 Ashdown, Inez, The Lahaina District, pp. 15, 16 (MHS) A15 7-7B.

59 Ashdown, Inez, The Lahaina District, The Olowalu Massacre, p. 13 (MHS) A15 7-7B.

60 Ashdown, Inez, *Ke Alaloa O Maui*, p. 62, Kama'āina Historians, Inc., 1971.

61 Ashdown, Inez, Ali'i Nī'au Pi'o; Queen Ke'ōpūolani (MHS) A15 5-13.

62 Ashdown, Inez, Ali'i Nī'au Pi'o; Queen Ke'ōpūolani (MHS) A15 5-13.

63 Ashdown, Inez, *Ke Alaloa o Maui*, p. 46, Kama'āina Historians, Inc. 1971.

64 Speakman, Jr., Cummins, *Mowee,* p. 54, Pueo Press, 1978.

65 Speakman, Jr., Cummins, *Mowee,* p. 57, Pueo Press, 1978; Klieger, Christiaan, *Moku'ula*, p. 21, Bishop Museum Press, 1998.

66 Ashdown, Inez, Keawaiki, Sept 1981, p. 1, personal collection.

67 Ashdown, Inez, 'Āina Kaulana, Memoirs of Inez Ashdown, 1986, as told to Jill Engledow/Lynne Ashdown, p. 22, personal collection, see also (MHS) AR11 A2 1-1.

68 Ashdown, Inez, Ashdown for Jan Dapitan, Trails, p. 3, personal collection.

69 Ashdown, Inez, Towns of Maui, p. 1, (MHS) A15, 17-2.

70 Ashdown, Inez, Towns of Maui, p. 1, (MHS) A15, 17-2.

71 Ashdown, Inez, Beaches, personal collection.

72 Ashdown, Inez, Lahaina Town, (MHS) A15 10-26; The Twin Currents of the Eight Seas, personal collection.

73 Ashdown, Inez, A Tale of Old Lahaina, 4/26/72, (MHS) A15 11-8.

74 C. S. Stewart, *Journal of a Residence in the Sandwich Islands* 1823–1825, pp. 177, 182.

75 Ashdown, Inez, Mokuhinia Pond of Luaʻehu, (MHS) AR15 10-27.

76 Ashdown, Inez, handwritten note by Mrs. Ashdown in her personal copy of *The Story of Lahaina*, May 1947.

77 Ashdown, Inez, Luaʻehu makai, 1/22/76, pp. 1, 2 (MHS) A15 10-26.

78 Ashdown, Inez, Luaʻehu makai, 1/22/76, pp. 1, 2 (MHS) A15 10-26.

79 Ashdown, Inez, Luaʻehu area, (MHS) A15 10-27; Mokuhinia Pond of Luaʻehu, (MHS) A15 10-27.

80 Ashdown, Inez, Lahaina History, 9/24/75, p. 1, (MHS) A15 10-8; Beckwith, Martha, *Hawaiian Mythology*, pp. 393, 394, University of Hawaii Press, 1976.

81 Ashdown, Inez, The Lahaina District, Mokuhinia (MHS) A15 7-7A; The Kiha, (MHS) A15 10-1.

82 Ashdown, Inez, Luaʻehu area, (MHS) A15 10-27.

83 Ashdown, Inez, Mokuhinia Pond, p. 1, (MHS) A15 10-27.

84 Jensen, Lucia Tarallo and Natalie Mahina, *Daughters of Haumea*, p. 156, Pueo Press, 2005.

85 Klieger, C. P., *Mokuʻula*, p. 8, Bishop Museum Press, 1998.

86 Ashdown, Inez, *Ke Alaloa o Maui*, pp. 59, 60, Kamaʻāina Historians, Inc., 1971.

87 Kamakau, Samuel, *People of Old*, p. 83.

88 Ashdown, Inez, Luaʻehu area, (MHS) A15 10-27.

89 Ashdown, Inez, Mokuhinia Pond, p. 3, (MHS) A15 10-27.

90 Ashdown, Inez, Mokuhinia, (MHS) A15 7-7A.

91 Ashdown, Inez, Luaʻehu, 4/3/82, personal collection.

92 Ashdown, Inez, The Kiha, (MHS) A15 10-1.

93 Jensen, Lucia Tarallo and Natalie Mahina, *Daughters of Haumea*, Akua Moʻo, pp. 155–159, Pueo Press, 2005.

94 Ashdown, Inez, Mokuhinia, (MHS) A15 7-7A.

95 Klieger, C. P., *Mokuʻula*, Figure 6, p. 16, Bishop Museum Press, 1998.

96 Ashdown, Inez, *The Story of Lahaina*, pp. 29, 30, Taylor Publishing, 1947.

97 Klieger, C. P., *Mokuʻula*, p. 94, Bishop Museum Press, 1998.

98 Ashdown, Inez, Mokuhinia (MHS) A15 7-7A.

99 Klieger, C. P., *Mokuʻula*, p. 92, Bishop Museum Press, 1998.

100 Ashdown, Inez, Chiefs, (MHS) AR11 1-24.

101 Klieger, C. P., *Mokuʻula*, pp. 97, 98, Bishop Museum Press, 1998.

102 Klieger, C. P., *Moku'ula*, p. 101, Bishop Museum Press, 1998.

103 Ashdown, Inez, Lahaina District, The Hauola Stone, (MHS) A15 10-7; The Laha'āina District (MHS) A15 7-7A; Hauola Stone, (MHS) A15 10-26.

104 Ashdown, Inez, More about Heiau (Temples), p. 4, (MHS) A15 6-9.

105 Speakman, Jr., Cummins, *Mowee,* pp. 95, 96, Pueo Press, 1978.

106 Ashdown, Inez, Letter to Ray Morris, 12/27/63, (MHS) A15 10-2.

107 Klieger, C. P., *Moku'ula*, pp. 22, 23, Bishop Museum Press, 1998.

108 Ashdown, Inez, Background of Lele, p. 2, (MHS) A15 10-26; *Kamehameha-nui, King of Maui,* 3/24/67, Maui Revisited, personal collection.

109 Ashdown, Inez, Lahaina side, 6/10/75, pp. 8, 9, (MHS) A15 10-9.

110 Klieger, C. P., *Moku'ula*, p. 23, Bishop Museum Press, 1998.

111 The History of the Sandalwood Trade in Hawai'i, National Park Service @ cr.nps.gov/history.

112 Klieger, C. P., *Moku'ula*, pp. 22, 23, Bishop Museum Press, 1998.

113 U.S. Census @kapakulture.org.

114 Speakman, Jr., Cummins, *Mowee,* p. 79, Pueo Press, 1978.

115 Ashdown, Inez, Lahaina District, p. 25, (MHS) A15 7-7A.

116 Ashdown, Inez, General History, p. 1, (MHS) A15 10-8.

117 Klieger, C. P., *Moku'ula*, pp. 22, 23, Bishop Museum Press, 1998.

118 Ashdown, Inez, Lahaina District, p. 21, (MHS) A15 7-7A; Speakman, Jr., Cummins, *Mowee,* p. 98, Pueo Press, 1978.

119 The Henry 'Ōpūkaha'ia Collection, www.obookiah.com.

120 Speakman, Jr., Cummins, *Mowee,* p. 80, Pueo Press, 1978.

121 Speakman, Jr., Cummins, *Mowee,* p. 81, Pueo Press, 1978.

122 Alexander, W. D., *A Brief History of the Hawaiian People*, p 180, American Book Co., 1891.

123 Ashdown, Inez, Story of 'E'eke, 1980, p 4, personal collection.

124 Klieger, C. P., *Moku'ula,* p. 27, Bishop Museum Press, 1998.

125 Klieger, C. P., *Moku'ula*, p. 29, Bishop Museum Press, 1998.

126 Klieger, C. P., *Moku'ula*, pp. 29, 30, Bishop Museum Press, 1998.

127 Klieger, C. P., *Moku'ula,* p. 33, Bishop Museum Press, 1998.

128 Speakman, Jr., Cummins, *Mowee,* p. 81, Pueo Press, 1978.

129 Speakman, Jr., Cummins, *Mowee,* p. 87, Pueo Press, 1978.

130 Klieger, C. P., *Moku'ula*, p. 33, Bishop Museum Press, 1998.

131 Ashdown, Inez, The Lahaina District, The Mission Houses, (MHS) A15 7-7A; General History, 4/21/75, (MHS) A15 10-8.

132 Ashdown, Inez, General History, p. 2 (MHS) A15 10-8; The Kiha, p 1 (MHS) A15 10-1; Lahaina side, 6/10/75, p. 11, (MHS) A15 10-9; Commerce and Hawaii, Jan. 1982, p. 2, personal collection.

133 Ashdown, Inez, Lahaina District, p. 28, (MHS) A15 7-7A.

134 Ashdown, Inez, Waine'e. In Luaehu of Laha'āina, 6/30/75, personal collection.

135 Klieger, C. P., *Moku'ula*, p. 94, Bishop Museum Press, 1998.

136 Ashdown, Inez, Waine'e, 6/30/75, pp. 1, 2, personal collection; Protestant Mission Churches, (MHS) A15 7-7A.

137 Ashdown, Inez, Lahaina side, 6/10/75, p. 12, (MHS) A15 10-9.

138 Ashdown, Inez, *Stories of Old Lahaina*, 7th edition, p. 80, 1976, Hawaiian Service, Inc.

139 Lahaina Restoration Foundation, *Mo'olelo o Lahaina* brochure.

140 Ashdown, Inez, *Stories of Old Lahaina*, 7th edition, p. 7, 1976, Hawaiian Service, Inc.

141 Ashdown, Inez, The Lahaina District, The Mission Houses, p. 29, (MHS) A15 7-7A.

142 Ashdown, Inez, Plans for Lahaina Community, Aug. 1949, (MHS) A15 10-1.

143 Ashdown, Inez, The Lahaina District, Lahainaluna, pp. 16, 17, (MHS) A15 7-7B.

144 Ashdown, Inez, The Lahaina District, Lahainaluna, pp. 16, 17, (MHS) A15 7-7B.

145 S. M. Kamakau, *Ruling Chiefs of Hawaii*, p. 420; Ashdown, Inez, *The Story of Lahaina*, 1947, p. 15.

146 Taylor, Lois, *After 136 years, This Building Needs Help,* Star-Bulletin & Advertisers, April 20, 1980.

147 Speakman, Jr., Cummins, *Mowee,* pp. 105, 106, Pueo Press, 1978.

148 Ashdown, Inez, Fourth Lesson, Mauiana, p. 3, 10/9/57 (MHS) 2-13; Lua'ehu area, p. 1 (MHS) A15 10-27.

149 Maly, Kepa & Onaona, A Collection of Traditions and Historical Accounts of Kau'ula and other lands of Lahaina, Maui, 2007, Summary of Findings, pp. 4, 6.

150 Ashdown, Inez, The Lahaina District, pp. 16, 17, (MHS) A15 7-7B.

151 Taylor, Lois, *After 136 years, This Building Needs Help*, Star-Bulletin & Advertisers, April 20, 1980.

152 Speakman, Jr., Cummins, *Mowee,* p. 82, Pueo Press, 1978.

153 Ashdown, Inez, General History, p. 2, (MHS) A15 10-8; Keawaiki, p. 2, (MHS) A15 10-26.

154 Ashdown, Inez, Lahaina District, p. 23, (MHS) A15 7-7A.

155 Klieger, C. P,. *Moku'ula*, pp. 42, 43, Bishop Museum Press, 1998.

156 Speakman, Jr., Cummins, *Mowee,* pp. 88, 89, Pueo Press, 1978.

157 Klieger, C. P., *Moku'ula*, p. 43, Bishop Museum Press, 1998.

158 Speakman, Jr., Cummins, *Mowee,* p. 91, Pueo Press, 1978.

159 Klieger, C. P., *Moku'ula*, p. 44, Bishop Museum Press, 1998.

160 Klieger, C. P., *Moku'ula*, p. 46, Bishop Museum Press, 1998.

161 Ashdown, Inez, Lahaina Town, p. 30 (MHS) A15 7-7A; Frost, H.l. & R.M, *Report on Seamans Hospital*, Maui News, 12/31/75.

162 Ashdown, Inez, The Mission Houses, (MHS) A15 7-7A; Lahaina side, p. 14 (MHS) A15 10-9; General History, p. 2, (MHS) A15 10-8.

163 Ashdown, Inez, Note cards for Herb Kane, Heiau of Ka'anapali, personal collection.

164 Klieger, C. P., *Moku'ula*, pp. 52–53, Bishop Museum Press, 1998.

165 Alexander, W. D., *A Brief History of the Hawaiian People*, p. 230, American Book Co., 1891.

166 Klieger, C. P., *Moku'ula*, p. 46, Bishop Museum Press, 1998.

167 Van Dyke, Jon M., *Who Owns the Crown Lands of Hawaii?,* pp. 44–49, 2007, UH Press.

168 Ashdown, Inez, General History, 4/21/75, (MHS) A15 10-8.

169 Ashdown, Inez, Crime and Punishment in Lahaina, 4/21/75, (MHS) A15 10-26.

170 Ashdown, Inez, Lahaina Town, p. 21, (MHS) A15 10-26.

171 Ashdown, Inez, Lahaina Town, p. 22, (MHS) A15 10-26.

172 Ashdown, Inez, Lahaina District, (MHS) A15 7-7A; Lahaina Town, pp. 20, 21, (MHS) A15 10-26.

173 Ashdown, Inez, *Stories of Old Lahaina*, 7th edition, p. 24, Hawaiian Service, Inc., 1976.

174 Ashdown, Inez, Historical Sites of Maui, 1957, (MHS) A15 10-8.

175 Ashdown, Inez, Notecard for Herb Kane, Heiau of Ka'anapali, personal collection.

176 Ashdown, Inez, Chiefs, p. 4, (MHS) AR11 1-24.

177 Ashdown, Inez, Lahaina Town, (MHS) A15 10-26.

178 Ashdown, Inez, Sugar and Pineapple in Lahaina District, (MHS) A15 7-7D; olowalu.net.

179 Ashdown, Inez, Maui Historical Society, Historic Sites of West Maui,2/17/57, (MHS) A15 10-8.

180 Ashdown, Inez, Lahaina District, p. 31, (MHS) A15 7-7A.

181 Ashdown, Inez, Lahaina District, p. 32, (MHS) A15 7-7A; www.lahainarestoration.org.

182 Ashdown, Inez, Waine'e in Lua'ehu of Lahaina, 6/30/75, personal collection.

183 Japanese Buddhist Temples @ Hawaii.gov.

184 Speakman, Jr., Cummins, *Mowee,* p. 148, Pueo Press, 1978.

185 Lahaina Jodo Mission, www.lahainajodomission.org.

186 Ashdown, Inez, To Herb Kane, personal collection.

187 Ashdown, Inez, For Pua Lindsey & family and the Hui Wa'a o Maui, 4/5/80, p. 9, pers. coll.

188 Ashdown, Inez, Priority Historic Sites List for Registry of Historic Places, 5/5/76, (MHS) A15 2-1.

189 Ashdown, Inez, Pōhaku Pe'e, personal collection; Lahaina Area, 2/25/75, pp. 1, 2, (MHS) A15 11-6A.

190 Ashdown, Inez, Ka'anāpali, Home of Dieties, 9/67, (MHS) A15 9-6.

191 Ashdown, Inez, Lahaina District, Mala, pp. 33–36, (MHS) A15 7-7A.

192 Ashdown, Inez, Lahaina District, p. 33, (MHS) A15 7-7A.

193 Ashdown, Inez, Priority Historic Sites List for West Maui, 5/5/76, p. 1, (MHS) A15 2-1.

194 Ashdown, Inez, Lahaina District, p. 36, (MHS) A15 7-7A.

195 Ashdown, Inez, Ka'anapali notes for Royal Lahaina Hotel, Pilahi Paki, Herb Kane, Amfac, 1980, personal collection.

196 Ashdown, Inez, The Battle of Kokoonāmoku, (MHS) A15 10-8.

197 Ashdown, Inez, Lahaina District, pp. 14, 15, (MHS) A15 7-7B.

198 Ashdown, Inez, Lahaina Town, (MHS) A15 10-26.

199 Ashdown, Inez, Kalana O Maui, Lahaina, pp. 63, 64, personal collection.

200 Ashdown, Inez, *Kealaloa o Maui*, p. 17; Lahaina District, p. 40, (MHS) A15 7-7A.

201 Ashdown, Inez, Lahaina Area 2/25/75, p. 3 (MHS) A15 11-6A.

202 Ashdown, Inez, For 'Īao Valley Lodge, 5/15/74, personal collection.

203 Ashdown, Inez, Iao, Sacred Valley of Worthy Ali'i, personal collection.

204 Ashdown, Inez, Runners, for Darby Gill, p. 3, personal collection.

205 Nakuina, E. M., *A Legend of Kanikaniaula and the First Feather Cloak.*

206 Ashdown, Inez, The Time of the Ali'i, Part II, p. 12, (MHS) AR11 1-23.

207 Nakuina, E. M., *A Legend of Kanikaniaula and the First Feather Cloak, Hawai-*

ian Folk Tales: A Collection of Native Legends, p. 155; Brigham, Wm. T., *Hawaiian Featherwork,* p. 59, *Memoirs of Bishop Museum, Vol. 1, No. 1, 1899.*

208 Ashdown, Inez, Notes, chiefs, p. 3, (MHS) A15 5-13.

209 Ashdown, Inez, Lahaina District, p. 43, (MHS) A15 7-7A.

210 Ashdown, Inez, Lāna'i The Bouyant, 10/14/73, personal collection.

211 Ashdown, Inez, Lahaina Area, 2/26/75, p. 3, (MHS) A15 11-6A; History of Honolua, personal collection, see also (MHS) AR11 A2 2-60.

212 Ashdown Inez, *Kealaloa o Maui,* p. 46, 1971, Kama'aina Historians, Inc.

213 Ashdown, Inez, For Herb Kane, Index card no.6, personal collection; Pōhaku Ka'anāpali, (MHS) A15 9-6, Historic Sites of West Maui, 2/17/57, (MHS) A15 10-8.

214 Ashdown, Inez, Leap of Death Games, 11/9/63; Maui's Last Great King, 11/6/66; For Herb Kane, Pu'u Keka'a, p. 6, personal collection.

215 Ashdown, Inez, Lahaina Area, 2/25/75, p. 3, (MHS) A15 11-6A; Pōhaku Moemoe, personal collection, Gods of Maui, 9/62, personal collection.

216 Ashdown, Inez, Lahaina District, pp. 39, 40, (MHS) A15 7-7A.

217 Ashdown, Inez, Pele on Moku Maui, personal collection; Realm of the Gods, p. 5, (MHS) A15 11-10.

218 Ashdown, Inez, The Bays or Hono of the Lahaina Coast, (MHS) A15 7-7B; Ka'anāpali Now, p. 3, personal collection; History of Honolua Ranch, p. 5, personal collection, see also (MHS) AR11 A2 2-60.

219 Ashdown, Inez, The Bays or Hono of Lahaina Coast, pp. 42–49, (MHS) A15 7-7A.

220 Ashdown, Inez, The War Hiding Caves of Maui, personal collection; Lahaina side, 6/16/75, (MHS) A15 10-9.

221 Ashdown, Inez, History of Honolua Ranch, p. 4, pers. coll., see also (MHS) AR11 A2 2-60.

222 Ashdown, Inez, The Lahaina District, pp. 38, 39 (MHS) A15 7-7A, as told by Alice Kaehukai Ka'ae and Lahela Reimann.

223 Ashdown, Inez, Beaches, personal collection.

224 Ashdown, Inez, The Bays or Hono of Lahaina Coast, p. 43, (MHS) A15 7-7A, as related by Lahela Reimann.

225 Ashdown, Inez, The Lahaina District, pp. 41, 42 (MHS) A15 7-7A.

226 Ashdown, Inez, History of Honolua Ranch, p. 6, pers. coll., see also (MHS) AR11 A2 2-60.

227 M. Pietrusewsky, *Human skeletal and dental remains from the Honokahua burial site, Land of Honokahua*, 1991.

228 Ashdown, Inez, Phyllis Fox, Historic Hawaii Foundation, Heiau sites of Lahaina, 11/30/77 (MHS) A15 10-7;History of Honolua Ranch, p. 7, pers. coll., also (MHS) AR11 A2 2-60.

229 Ashdown, Inez, History of Honolua Ranch, p. 6, pers. coll., also (MHS) AR11 A2 2-60.

230 Ashdown, Inez, History of Honolua Ranch, p. 8, pers. coll., also (MHS) AR11 A2 2-60.

231 Ashdown, Inez, Lahaina assignment, June 1975, James Young Kanehoa (MHS) AR 11 8-14.

232 Ashdown, Inez, Lahaina assignment, June 1975, James Young Kanehoa (MHS) AR 11 8-14.

233 Hawaii State Archives, James Young Kanehoa.

234 Ashdown, Inez, The History of Honolua Ranch, p. 17, pers. coll., also (MHS) AR11 A2 2-60.

235 Ashdown, Inez, History of Honolua, p. 32, pers. coll., ; Battlefields, Puʻu Kahuahua (MHS) AR11 1-43; Historic Sites List, Historic Sites Commission, 9/29/56, (MHS) AR11 1-31.

236 Ashdown, Inez, Lahaina Assignment, June 1975, (MHS) A15 8-14; History of Honolua Ranch, p. 32, personal collection, also (MHS) AR11 A2 2-60.

237 Ashdown, Inez, The History of Honolua Ranch, p. 13, pers. coll., also (MHS) AR11 A2 2-60.

238 Ashdown, Inez, from Hattie Kanaka-o-kai Yoshikawa, note card, personal collection.

239 Ashdown, Inez, History of Honolua Ranch, p. 29, pers. coll., also (MHS) AR11 A2 2-60.

240 Ashdown, Inez, History of Honolua Ranch, pp. 32, 33, pers. coll., also (MHS) AR11 A2 2-60.

241 Ashdown, Inez, History of Honolua Ranch, p. 36, pers. coll., also (MHS) AR11 A2 2-60.

242 Ashdown, Inez, History of Honolua Ranch, p. 35 pers. coll., also (MHS) AR11 A2 2-60.

243 History of D. T. Fleming and the Fleming Arboretum, @flemingarboretum.org.

244 Ashdown, Inez, History of Honolua Ranch, p. 36, pers. coll., also (MHS) AR11 A2 2-60.

245 Ashdown, Inez, History of Honolua Ranch, p. 37, pers. coll., also (MHS) AR11 A2 2-60.

246 Ashdown, Inez, History of Honolua Ranch, p. 38, pers. coll., also (MHS) AR11 A2 2-60.

247 History of D. T. Fleming and the Fleming Arboretum, @flemingarboretum.org; Ashdown, Inez, History of Honolua Ranch, pp. 34–36, pers. coll., also (MHS) AR11 A2 2-60.

248 History of D. T. Fleming and the Fleming Arboretum, @flemingarboretum.org.

249 Ashdown, Inez, Sugar and Pineapple in Lahaina Dist., (MHS) A15 7-7D; History of Honolua Ranch, personal collection, also (MHS) AR11 A2 2-60.

250 Maui News, *New Twist to dilemma over burial site*, 12/21/88.

251 Maui News, *Waihee: No pledge made to Kapalua*, 2/24/81.

252 M. Pietrusewsky, *Human skeletal and dental remains from the Honokahua burial site, Land of Honokahua*, 1991.

253 Maui News, *Burials prominent topic at archaeology meeting*, 4/3/89.

254 state.hi.us/dlnr/hpd/naiwikupuna.htm.

255 Ashdown, Inez, Ka Palaoa, The Ivory Symbol; The Clean Sands of Na Kai ʻEwalu, personal Collection, Hauola Stone, (MHS) AR 11 10-26.

256 Donham, Theresa, Honokahua survey, 1986, PHRI, Hawaii.

257 Ashdown, Inez, *Kealaloa o Maui*, p. 5, Kamaʻaina Historians, Inc.

258 Walker, Winslow, *Archaeology of Maui*, 1931, p. 70, Bishop Museum.

259 Ashdown, Inez, The History of Honolua Ranch, p. 37, pers. coll., also (MHS) AR11 A2 2-60.

260 Ashdown, Inez, Lahaina District, p. 48 (MHS) A15 7-7A.

261 Save Honolua Coalition @savehonolua.org.

262 Walker, Winslow, *Archaeology of Maui*, 1931, site no.19, p. 72, Bishop Museum.

263 Walker, Winslow, *Archaeology of Maui*, 1931, site no. 20, p. 73, Bishop Museum.

264 Ashdown, Inez, History of Honolua Ranch, p. 1, pers. coll., also (MHS) AR11 A2 2-60.

265 Sterling, Elspeth, Sites of Maui, p. 47, Kaʻanapali District 5, 1998, Bishop Museum Press.

266 Ashdown, Inez, History of Honolua Ranch, p. 2, pers. coll., also (MHS) AR11 A2 2-60.

267 Sterling, Elspeth, Sites of Maui, p. 53, Honkohau 31, 1998, Bishop Museum Press.

268 Ashdown, Inez, History of Honolua Ranch, p. 10, pers. coll., also (MHS) AR11 A2 2-60.

269 Ashdown, Inez, History of Honolua Ranch, p. 2, pers. coll., also (MHS) AR11 A2 2-60.

270 Ashdown, Inez, History of Honolua Ranch, pp. 2, 3, pers. coll., also (MHS) AR11 A2 2-60.

271 Ashdown, Inez, Lahaina District, pp. 48, 49, (MHS) A15 7-7A.

272 nydailynews.com, *Calif. man sucked into blowhole*, 7/14/11, mauinow.com, 10/2/13.

273 Walker, Winslow, *Archaeology of Maui*, 1931, site no. 21, p. 74, Bishop Museum.

274 Ashdown, Inez, Ka'anāpali Now, p. 3, personal collection, see also (MHS) AR11 2-72.

275 Ashdown, Inez, Maui's Bell Stone, for Jan Dapitan, Parks Dept., (MHS) A15 11-6A; Religion, p. 2, (MHS) A15 15-25.

276 Ashdown, Inez, Agricultural pursuits at Ka'anāpali, personal collection.

277 S. Kamakau, Kuokoa, January 12, 1867.

278 Ashdown, Inez, Maui's Ancient Trails, personal collection.

279 Ashdown, Inez, *Ke Alaloa o Maui*, p. 60, 1971, Kama'aina Historians, Inc.

280 Walker, Winslow, *Archaeology of Maui*, 1931, site numbers 22, 23, 25, Bishop Museum.

281 Walker, Winslow, *Archaeology of Maui*, 1931, site number 24, Bishop Museum.

282 Ashdown, Inez, *Ke Alaloa o Maui*, p. 37, 1971, Kama'aina Historians, Inc.

283 Ashdown, Inez, *Ke Alaloa o Maui*, p. 37; Sterling, Elspeth, *Sites of Maui*, p. 58.

284 Speakman, Jr., Cummins, *Mowee*, p. 2, Pueo Press, 1978.

285 Beckwith, Martha, *Hawaiian Mythology*, p. 226, University of Hawaii Press, 1970.

286 Ashdown, Inez, Ulu Genealogy, 1966, (MHS) A15 5-13.

287 Liliu'okalani of Hawaii, *The Kumulipo*, pp. 77, 78, Pueo Press, 1978.

288 Ashdown, Inez, *The Old Trails of Moku Maui*, The Valley Isle, 6/28/63.

289 Lili'uokalani of Hawaii, *The Kumulipo*, p. 76, Pueo Press, 1978.

290 Ashdown, Inez, The Time of the Ali'i, Part Two, pp. 7, 8, (MHS) AR11 1-23.

291 Kaha, S, 1872, Fornander Collection of Hawaiian Folklore, Pukui, 1983.

292 Ashdown, Inez, *The Old Trails of Moku Maui*, The Valley Isle, June 28, 1963.

293 Ashdown, Inez, Ulu Genealogy, 1966, (MHS) A15 5-13.

294 Ashdown, Inez, 'Āina Kaulana, Memoirs of Inez Ashdown, 1986, as told to Jill Engledow/Lynne Ashdown, p. 1, personal collection, see also (MHS) AR11 A2 1-1.

295 Ashdown, Inez, *A Rodeo, A President, and Hawaiians*, Oct. 19, 1963, *The Maui News*.

296 Ashdown, Inez, *A Child Meets A Queen*, Oct. 26, 1963, *The Maui News.*

297 Ashdown, Inez, ʻĀina Kaulana, Memoirs of Inez Ashdown, 1986, as told to Jill Engledow/Lynne Ashdown, p. 17, personal collection, see also (MHS) AR11 A2 1-1.

298 Low, Eben, Eben Parker Low Memoir, Chapter II, Family History, personal collection, also at MHS A15 12-1; Klieger, C. P., *Mokuʻula*, p. 68, Bishop Museum Press, 1998.

299 Ashdown, Inez, ʻĀina Kaulana, Memoirs of Inez Ashdown,1986, as told to Jill Engledow/Lynne Ashdown, pp. 3, 4, personal collection, see also (MHS) AR11 A2 1-1.

300 Ashdown, Inez, *A Rodeo, A President and Hawaiians*, Oct. 19, 1963, *The Maui News.*

301 Ashdown, Inez, *Journey with a Queen and her Court*, Nov. 30, 1963, *The Maui News.*

302 Ashdown, Inez, *Farewell to my Queen and Welcome to Maui*, Dec. 7, 1963, *The Maui News.*

303 Ashdown Inez, ʻĀina Kaulana, Memoirs of Inez Ashdown,1986, as told to Jill Engledow/Lynne Ashdown, pp. 28, 29, personal collection, see also (MHS) AR11 A2 1-1.

304 Ashdown, Inez, Maui's Champion Woman, manuscript on Armine von Tempski, pers. coll., See also MHS A15 17-3 von Tempsky Family.

305 Ashdown, Inez, ʻĀina Kaulana, Memoirs of Inez Ashdown, 1986, as told to Jill Engledow/Lynne Ashdown, p. 55, personal collection, see also (MHS) AR11 A2 1-1.

306 Ashdown, Inez, ʻĀina Kaulana, Memoirs of Inez Ashdown, 1986, as told to Jill Engledow/Lynne Ashdown, pp. 57–61, personal collection, see also (MHS) AR11 A2 1-1.

307 Ashdown, Inez, ʻĀina Kaulana, Memoirs of Inez MacPhee Ashdown, as told to Jill Engledow/Lynne Ashdown, p. 86, personal collection, see also (MHS) AR11 A2 1-1.

308 von Tempski, Armine, *Born in Paradise*, p. 264,Ox Bow Press, reprint 1985.

309 Emory, Kenneth P. *An Archaeological Survey of Haleakala,* Bishop Museum Press, 1922.

310 *Ashdown is a "true woman paniolo," The Maui News,* Makawao Rodeo Parade Supplement, July, 1982.

311 Ashdown, Inez, *Recollections of Kahoʻolawe,* p. 41, Topgallant Publishing Co, 1979.

312 Ashdown, Inez,ʻĀina Kaulana, Memoirs of Inez MacPhee Ashdown, as told to

Jill Engledow/Lynne Ashdown, pp. 73, 80, personal collection, see also (MHS) AR11 A2 1-1.

313 Ashdown, Inez, ʻĀina Kaulana, Memoirs of Inez MacPhee Ashdown, as told to Jill Engledow/Lynne Ashdown, pp. 80, 82, 86, 89, 91, 92, personal collection, see also (MHS) AR11 A2 1-1.

314 Ashdown, Inez, ʻĀina Kaulana, Memoirs of Inez MacPhee Ashdown, as told to Jill Engledow/Lynne Ashdown, p. 96, personal collection, see also (MHS) AR11 A2 1-1.

315 Ashdown, Inez, ʻĀina Kaulana, Memoirs of Inez MacPhee Ashdown, as told to Jill Engledow/Lynne Ashdown, p. 98, personal collection, see also (MHS) AR11 A2 1-1.

316 Ashdown, Inez, ʻĀina Kaulana, Memoirs of Inez MacPhee Ashdown, as told to Jill Engledow/Lynne Ashdown, p. 125, personal collection, see also (MHS) AR11 A2 1-1,.

317 Ashdown, Inez, ʻĀina Kaulana, Memoirs of Inez MacPhee Ashdown, as told to Jill Engledow/Lynne Ashdown, pp. 97, 98, personal collection, see also (MHS) AR11 A2 1-1.

318 Klieger, C. P., *Mokuʻula*, p. 68, Bishop Museum Press, 1998.

319 Ashdown, Inez, The Shaws, 1/22/76, pp. 1, 2, (MHS) A15 15-3.

320 Ashdown, Inez, The Shaws, 1/22/76, pp. 1, 2, (MHS) A15 15-3.

321 Ashdown, Inez, ʻĀina Kaulana, Memoirs of Inez MacPhee Ashdown, as told to Jill Engledow/Lynne Ashdown, p. 98, personal collection, see also (MHS) AR11 A2 1-1.

322 Ashdown, Inez, Lahaina notes, (MHS) A15 10-1.

323 Ashdown, Inez, Lahaina history (MHS) A15 10-8.

324 Ashdown, Inez, Lahaina Town, p. 26, (MHS) A15 7-7A.

325 Ashdown, Inez, Letter to Pilahi Paki, Lahaina assignment, 5/8/75, (MHS) A15 11-6B.

326 Ashdown, Inez, ʻĀina Kaulana, Memoirs of Inez MacPhee Ashdown, as told to Jill Engledow/Lynne Ashdown, p. 99, personal collection, see also (MHS) AR11 A2 1-1.

327 Ashdown, Inez, ʻĀina Kaulana, Memoirs of Inez MacPhee Ashdown, as told to Jill Engledow/Lynne Ashdown, p. 101, personal collection, see also (MHS) AR11 A2 1-1.

328 Ashdown, Inez, ʻĀina Kaulana, Memoirs of Inez MacPhee Ashdown, as told to Jill Engledow/Lynne Ashdown, p. 102, personal collection, see also (MHS) AR11 A2 1-1.

329 Ashdown, Inez, *Kaho'olawe's Red Dust Desert Turned into Productive Domain*, p. 1, Third Section, *Honolulu Star Bulletin*, February 6, 1937.

330 Ashdown, Inez, 'Āina Kaulana, Memoirs of Inez MacPhee Ashdown, as told to Jill Engledow/Lynne Ashdown, p. 115, personal collection, see also (MHS) AR11 A2 1-1.

331 Ashdown, Inez, 'Āina Kaulana, Memoirs of Inez MacPhee Ashdown, as told to Jill Engledow/Lynne Ashdown, pp. 112, 113, 117, personal collection, see also (MHS) A2 1-1.

332 Ashdown, Incz, 'Āina Kaulana, Memoirs of Inez MacPhee Ashdown, as told to Jill Engledow/Lynne Ashdown, p. 125, personal collection, see also (MHS) AR11 A2 1-1.

333 Ashdown, Inez, Sacred Temple of Olowalu, 10/67, (MHS) A15 6-9.

334 Ashdown, Inez, 'Āina Kaulana, Memoirs of Inez MacPhee Ashdown, as told to Jill Engledow/Lynne Ashdown, p. 123, personal collection, see also (MHS) AR11 A2 1-1.

335 Ashdown, Inez, Wailuku Historic District, July, 1970, (MHS)A15 17-7.

336 *Bishop Museum work finds features at Wailea, recommends living museum, The Maui News*, 7/15/70, p. A12.

337 Cox, Richard, Wailea Development Company Letter to Mrs. Ashdown, 12/2/81, pers. coll.

338 *Stricken from State Register*, Historic Hawaii News, p. 1, April, 1980.

339 Ashdown, Inez, 'Āina Kaulana, Memoirs of Inez MacPhee Ashdown, as told to Jill Engledow/Lynne Ashdown, pp. 129, 130, personal collection, see also (MHS) AR11 A2 1-1.

340 King, Lieutenant Goveror Jean, Letter to Inez Ashdown, October 24, 1979,Fox, Phyllis G., Exe. Director, Historic Hawaii Foundation, Letter to Inez Ashdown, 11/2/79, personal collection.

341 Office of the Mayor, Press Release, February 25, 1980, *Mayor Asks Governor to Help Save Hale Pa'i.*

342 Ashdown, Inez, personal note in personal Hale Pa'i file, personal collection.

343 *Hale Pa'i Blessing is Held*, p. B1, *The Maui News,* Sept. 12,1980.

344 *Hale Pa'i Blessing is Held*, p. B4, *The Maui News,* Sept. 12,1980.

345 Hale Pa'i Dedication pamphlet, Dec. 17, 1982.

346 Ashdown, Inez, Letter to Ralston Nagata, Director, State Parks Department, 3/29/83; State Parks Dept. Letter to U.S. Army Battalion Commander, 4/6/83; personal collection.

347 Proclamation, Mayor Hannibal Tavares, March 9, 1982, personal collection.

BIBLIOGRAPHY

Alexander, W. D. *Brief History of the Hawaiian People*, American Book Company, 1891.

Ashdown, Inez. *The Story of Lahaina*, 1947.

———. *Ke Alaloa O Maui*, Kama'aina Historians, Inc, 1971.

———. *Recollections of Kaho'olawe*, Topgallant Publishing Co., 1979 Beckwith, Martha, The Kumulipo, The University Press of Hawaii, 1972.

Beckwith, Martha,.*Hawaiian Mythology*, University of Hawaii Press, 1970.

Duensing, Dawn E. *Maui In History, A Guide to Resources*, State Foundation on Culture and the Arts, 1998.

Jensen, Lucia Tarallo and Natalie Mahina Jensen. *Na Kaikamahine 'o Haumea*, Daughters of Haumea, Pueo Press, 2005.

Klieger, Christiaan P. *Moku'ula, Maui's Sacred Island*, Bishop Museum Press, 1998.

Liliuokalani. *The Kumulipo*, Lee and Shepard of Boston, 1897, Pueo Press, 1978.

———. *Hawaii's Story by Hawaii's Queen*, 1898, Mutual Publishing, 1990.

Pukui, Mary K. and S. Elbert. *Hawaiian Dictionary*, University of Hawaii Press, Honolulu, 1986.

Speakman, Jr., Cummins, E. *Mowee*, The Nimrod Press, Boston, 1978, Pueo Press, 1984.

Sterling, Elspeth P. *Sites of Maui*, Bishop Museum Press, 1998.

INDEX

I

J

K

N

O

P